Ultra-Wideband Radio Technology

Ultra-Wideband Radio Technology

Kazimierz Siwiak and Debra McKeown

Both of

TimeDerivative Inc.,
USA

John Wiley & Sons, Ltd

This publication is designed to provide accurate and authoritative information in regard to the
subject matter covered. It is sold on the understanding that the Publisher is not engaged in
rendering professional services. If professional advice or other expert assistance is required, the
services of a competent professional should be sought.

Other Wiley Editorial Offices

John Wiley & Sons Inc., 111 River Street, Hoboken, NJ 07030, USA

Jossey-Bass, 989 Market Street, San Francisco, CA 94103-1741, USA

Wiley-VCH Verlag GmbH, Boschstr. 12, D-69469 Weinheim, Germany

John Wiley & Sons Australia Ltd, 33 Park Road, Milton, Queensland 4064, Australia

John Wiley & Sons (Asia) Pte Ltd, 2 Clementi Loop #02-01, Jin Xing Distripark, Singapore 129809

John Wiley & Sons Canada Ltd, 22 Worcester Road, Etobicoke, Ontario, Canada M9W 1L1

Wiley also publishes its books in a variety of electronic formats. Some content that appears
in print may not be available in electronic books.

British Library Cataloguing in Publication Data

A catalogue record for this book is available from the British Library

ISBN 0-470-85931-8

Typeset in 11/13pt Times by Laserwords Private Limited, Chennai, India
Printed and bound in Great Britain by TJ International, Padstow, Cornwall
This book is printed on acid-free paper responsibly manufactured from sustainable forestry
in which at least two trees are planted for each one used for paper production.

Contents

... for Ken who grew weary of explaining UWB to strangers in grocery store parking lots... Now you can tell them, "Buy the book!"

Preface

In several years of presenting UWB technology to potential commercial customers, investors, and the scientific, engineering, academic community at large, it has become apparent that a comprehensive text is needed. The need occurs at two levels: (1) fundamental, technically accurate information devoid of specific technical and analytical details is needed by marketing managers, business developers, engineering managers, technology managers, potential investors, financial analysts, executive recruiters, technical writers, and technologists from other fields, and (2) specific technical and engineering information about UWB in sufficient detail is needed for seasoned technologists, engineers, scientists, academicians who want to understand the topic at an entry level. We especially encourage students to explore the simplicity of UWB technology experimentally (see <http://timederivative.com/pubs.htm>). We will periodically provide updated materials, including questions and exercises, at <http://timederivative.com/UWBbook.html>

Our vision, then, is almost of two different books in one: the first, a simple, high-level, conceptual discussion of UWB, and the second, a more detailed portion, focusing on scientific, mathematical, engineering aspects. There are many drawings to explain the technology and nice analogies that are understandable and based on common knowledge. We present material on two levels: a fundamental level for the "nontechnical" and a technically detailed but entry level for the "seasoned technical" individuals. One of us is a technical specialist with much experience in commercial radio technologies and in technical writing. Another is a technical training and media specialist as well as a technical writer and teacher with much experience in presenting and engaging an audience in technical material. Together we have effectively produced presentations, tutorials, and in-house training for a wide variety of audiences.

Ultra-wideband (UWB) has been among the most controversial technologies of modern times. Its applications seem endless, its capabilities miraculous, and yet it is very poorly understood, even by those closest to it. We attempted to convey clarity of thought in difficult and unusual technical concepts. We expect UWB to have an impressive impact on the way we live our lives in the future. This book, therefore, is an effort to bring the basics of the technology to a wider audience, beyond just technologists, so that numerous imaginations can be set to work developing and implementing innovative ideas to make all of our lives more convenient, connected, safe, efficient, and fun.

This book is designed to give a basic overview of the subject – ground the reader with a brief history, offer an understanding of the current regulations and standards that are being developed – and only then do we present the workings of the technology. The reader will come to see issues that are usually problematic for other wireless solutions can be beneficial to ultra-wideband. We explain how to create UWB signals in theory as well as in practice. Finally, we suggest some possible applications, which we hope will serve as a springboard for our readers who will no doubt apply their own creativity and dream up advancements that we have not even considered. We hope that this book will serve to inform and educate, as well as inspire our readers as much as the technology has inspired us. Although most of the early initiatives were in the United States, we attempted a global view of the subject. In fact, the book itself was produced partially in four continents and across the globe through the Internet.

We introduce in Chapter 1, the history of ultra-wideband. To appreciate any technology, any subject for that matter, one needs to appreciate its history. We can better understand how a technology operates if we can understand how it was developed and on what previous knowledge it is based. UWB has an interesting beginning, its first appearance being in the earliest spark-gap mechanisms. Until now the wide bandwidths of early radio could not be used effectively. From these modest, low-tech beginnings, we have come to understand the high points of why UWB needed to wait for other technologies before it could develop into the high-tech marvel of today. Having paid due respect to its humble start, we can move onto understanding the nature of the technology as it stands today.

The development of a technology is tempered as much by invention and innovation as by regulations, which we explore in Chapter 2: The Regulatory Climate. All technologies have their own set of properties and constraints placed on them by physics as well as by regulations, which deal with the human impact of technology. Government regulators place constraints on the way technologies can operate, so as to make coexistence more harmonious and also to ensure public safety and to foster economic benefit to society. The reader will learn that regulations can actually change the way a technology evolves, how it performs, and how it is likely to enter the mass marketplace. In the United States, regulations do not define UWB so much as they define the rules under which the UWB spectrum may be accessed. The rules are so broad that some conventional radio technologies might "masquerade" as UWB. We note, however, that the purpose of this book is not to explore conventional technologies, but rather to explore those UWB techniques that exploit the beneficial aspects of UWB. We will concentrate on UWB solutions that have UWB characteristics. The regulations surrounding ultra-wideband are still evolving, so in this chapter, we capture what is current in the field. However, the authors note that in considering UWB, regulators have changed their approaches to spectrum management.

A technology becomes pervasive in society when it can be introduced in a standard fashion and can benefit from economies on a global scale. In Chapter 3: UWB in Standards, we trace the development of some standards activities in which UWB will appear. In addition to regulating how technologies interact with the rest of the world, we must also agree on how they will interact with each other and how multiple equipment suppliers can build UWB devices that operate seamlessly among different manufacturers. The standards picture, like the rest of UWB's history, is slow to develop and, at this writing, is still changing. Although standards are not critical to understand how a technology works or how it will perform, they are crucial if one wishes to develop an actual device for the marketplace. Therefore, in this chapter, we have laid out several of the options that are, at the time of printing, being proposed as *the* commercial standard. Any "standard" product development, of course, must await the definition of a standard.

The generation of wideband signals requires some different techniques than those used with conventional radio signals, and these are detailed in Chapter 4: Generating and Transmitting UWB Signals. The first step in communicating wirelessly is creating a suitable signal modulated with desired data to send to a destination. UWB offers some unique challenges in the generation stage, some of which are constraints imposed by regulators, and some are constraints imposed by physics. The reader will learn how to craft a UWB signal so that it fits within the constraints. Again, we concentrate on generating signals having UWB characteristics, since there are many works available to describe conventional radio technologies.

Radiation and propagation are described separately because interesting UWB properties are exposed in each of these processes. In Chapter 5: Radiation of UWB Signals, we see that the concept of *finite time* imparts interesting characteristics to UWB signal radiation. Wideband impulses are "there and gone," compared with the relatively long persistence of narrowband signals. This highly intermittent time nature gives UWB some of its unique and interesting radiation properties. We present the time solution to radiation and show how wideband signals differ from their narrowband counterparts.

Once radiated, the UWB signal traverses a path described in Chapter 6: Propagation of UWB Signals. A graceful information-rich signal is nice, but the intent is for it to be received. Before it can be received, though, it must first make its journey through our environment. Our homes and offices can be a dangerous place for signals, and in this chapter, we explain how UWB signals interact in the real world in a variety of environments. We discuss how the signals behave in various environments so that it may be understood how the unique properties of UWB propagation might be best exploited.

Emitted signals, of course, need to be received. In Chapter 7: Receiving UWB Signals, we capture the signal energy and retrieve the information that it carries. Receiving the UWB signal is an actual requirement under the UWB regulations! Receiving UWB signals is not very different from receiving other wireless signals. However, there is an art to receiving signals efficiently and translating them correctly to extract the information conveyed. In this chapter, we discuss the techniques of efficient signal reception.

One touted advantage of UWB is its enormous capacity. In Chapter 8: UWB System Limits and Capacity, we learn how the environment and other wireless users have an impact on the amount of information we can pack on a link. So, once we understand how to create, propagate, and receive signals efficiently, we must ask how many can share the medium. In this chapter, we learn to estimate performance and create link budgets. It is here that we tie the proverbial bow around the package – this is the chapter in which we *quantify* the performance of UWB wireless links.

We gaze into the crystal ball in Chapter 9: Applications and Future Directions. Some people will want to understand how the technology works, simply out of curiosity. It is our hope, though, that some readers will be inspired to continue their research and ultimately develop inventive products to enhance our world. This chapter explains several ideas that have been proposed for marketing and a couple that are just sparks of thought and are meant to enthuse readers and open minds to the possibilities.

Technology is not the sole domain of the "technologically advanced" segments of our global society. Although UWB appeared first in the United States, and the regulations permitting it was first penned there, we expect that the technology will have global impact. The Standards process is geared to make UWB available and cost effective everywhere. Our world is a complex mechanism, often made more complicated by technology. Technology tends to develop and advance initially in the developed and economically stable populations of the world, usually because of the massive expense of development. However, its deployment will not be thusly constrained. Cellular and mobile phones, for example, totally dominated places where landline telephones were nonexistent or just too expensive to lay out – the mobile phone growth in those global regions far outpaced that in technologically advanced places like the United States where the handheld cell phone was born. This is because the telephone system in the United States is a pretty system, while in the less technologically advanced areas of the world, mobile phones simply bypassed the expensive development of the older wire-line technology. UWB is likely to show up that way too. Many applications already have solutions that might not be perfect, but are good enough. Those places where no solutions exist right now are the most fertile for UWB. In the United States, it is "what can UWB do better?" In Kenya, it just might be "what can UWB do for which I have no solution at the moment!" UWB is seen as a low-cost solution to many communications problems today, which could make technology massively available in the developing nations. We encourage you to use your imagination to make this happen.

Kazimierz Siwiak and Debra McKeown

Acknowledgements

This work is a product of teamwork, by design. The team extends well past the synergy of the two named authors, though. Therefore, we thank many people for their generous contributions and efforts to make this work as up-to-date and accurate as is possible with such a fluid topic.

Thank you, Jay Bain for all your support regarding standards; Larry Fullerton for your assistance with the history...people tend to discuss the brilliance of your mind, but we would like to acknowledge the brilliance of your humanity as well; Paul Withington for all the technical discussions to clarify subtle points; Hans Schantz for always sharing your boundless electromagnetic knowledge; Laura Huckabee-Jennings for all of your help in the area of business direction and for your help when this book was still a tutorial.

Thanks to Gammz who patiently drew all of our silly ideas from stick figure sketches and who added his own flavor to everything with a style unique to Kenya.

Most of all, thanks to Kai for his vast knowledge, gentle nature, and open mind in response to my persistent urging to explain, define, and teach with examples...I'm sure every word he writes in the future will be burdened with my nagging voice in his head! – d.

... and likewise to Deb for her kindness, brilliance, and her tolerance of my retreats into techno-babble ... and for elevating the text with sparkle and clarity. – k.

This work is one produced on the road and over great distances, which can be very trying. There was a single night, though, when it all came together, so we must thank...

... Tori Amos and one glorious night in California...who inspired and healed, distracted, and focused, brought out every emotion binding the process...and put them back in order so we could continue, renewed...Thank you for using your voice to inspire us to find our own.

...and to Kenya whose perfect sun, vivid colors, and wild nature provided the setting for the final stages of the project.

Personally, I would like to acknowledge the influence of my ever-inspiring grandmothers Helen Boling McCoy and Joyce McKeown Owens for *living* the ethic of hard work and for personifying the essence of creativity. – d.

I would like to thank my family, Ann, Diana, and Joseph, for their support during this project and for walking with me on the journey we call life. Your love gives me purpose, strength and inspiration. – k.

1

History

Introduction

UWB – ultra-wideband – is an unconventional type of radio, but to understand a variation on the convention, we must grasp the basics of *traditional* radio. When most people hear the word "radio" they think of the small device that brings music and news into their homes and automobiles. That is true, but radio has many forms. In fact, many common devices that perform some function in a wireless mode are a variety of radio, such as wireless baby monitors, wireless Internet connections, garage door openers, and mobile or cell phones.

In this chapter, we introduce the basics of *traditional* radio. We follow the history of the development of wireless – to be dubbed *radio* at the start of the broadcast era – from its inception as crude wideband spark signals, through its relentless march towards narrowband-channelized solutions. Finally, we see its resurgence as the modern "wireless." History reveals that the march towards narrowband admits several instances in which wideband signaling has significant advantages over narrowband techniques. The present evolution to UWB is but an inevitable step in the evolution of wireless and radio.

1.1 The Basics of Radio

Radio is the art of sending and receiving electromagnetic signals between transmitters and receivers wirelessly, as depicted in Figure 1.1. Radio requires transmitters for generating signals, and receivers to translate the received information. Both use antennas for sending the signals as

Ultra-Wideband Radio Technology   Kazimierz Siwiak and Debra McKeown
© 2004 John Wiley & Sons, Ltd   ISBN: 0-470-85931-8

Figure 1.1 A basic radio link includes a transmitter, waves propagating and filling space, and a receiver [McKeown 2003].

electromagnetic energy and for collecting that energy at the receiver. Information, such as voice into a microphone, is supplied to transmitters, which then encode, or *modulate*, the information in some fashion on the signal. This information could be someone's voice, music, data, or other information. Receivers recover that information by decoding, or *demodulating*, the received signal and presenting it as received information.

Signals, electromagnetic energy-bearing information, inundate our surroundings. They usually originate from commercial broadcasting such as our familiar AM- and FM-band radio stations, television stations or consumer devices such as mobile phones, and garage door openers. There is a plethora of services that carry voice, music, video, telephony, and control instructions. There are also signals that originate from beyond the earth's immediate vicinity. They are natural stellar sources, pulsars, and such. Their "information" is carefully deciphered by radio astronomers to glean knowledge about our universe.

All signals, regardless of origin, simultaneously share the same "transmission medium" – the near vacuum of space, the air enveloping the earth and the many materials surrounding us. Yet we can selectively choose the signal we want, such as the radio station to which we wish to listen, the television program we want to watch, or the call intended for our mobile telephone. Radio signals in the electromagnetic spectrum (see Figure 1.2) keep us informed, entertained, and safe.

Conventional radio signals can be discerned one from the other because they occupy unique locations in the radio spectrum (see Figure 1.3), for

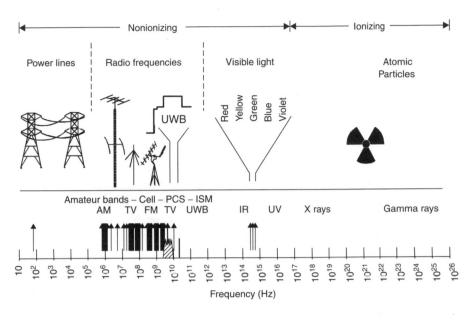

Figure 1.2 Radio services occupy unique locations in the electromagnetic spectrum.

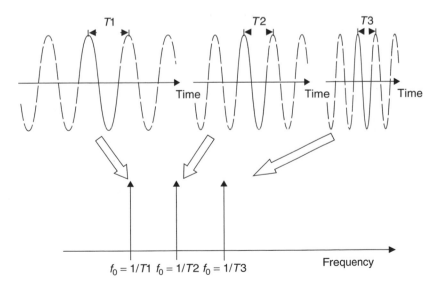

Figure 1.3 Sines and cosines of different wavelengths occupy unique spots in the spectrum.

instance, unique audio tones or discrete colors in the rainbow spectrum. They are distinct, narrowband places on a radio dial, indicated by channel numbers. They are all crafted that way because of a century-old historical interplay between the technological development of radio and the regulations that brought order to the radio spectrum. Radio signals share the limited spectrum by occupying slivers of spectrum that are as narrow as possible. A signal without information has zero bandwidth. Modulating information on that signal spreads its bandwidth in proportion to the information bandwidth. For example, a music signal with tonal content up to 15 kHz requires at least 15 kHz of information bandwidth. The "ideal" in radio spectrum usage has been to use the smallest bandwidth compared to the bandwidth of the signal information. Narrowband signals are often represented by their zero bandwidth ideals, the sine and cosine functions, also known as circular functions or harmonic waves. They are the narrowest possible representation of signals in the spectrum at distinct frequencies. Tuning radios to a particular frequency allows us to select the desired narrow band signal. So, the dogma of the circular functions, sines and cosines, [Harmuth 1968] began to dominate radio development.

Separation of signals by bands, by channels, and by frequencies is not the only way to share the radio spectrum. Information-bearing signals can also be separated in time, especially in tiny slivers of time. These

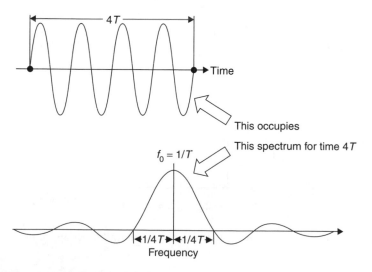

Figure 1.4   A finite length signal in time occupies a definite spectrum width in frequency, for that finite time.

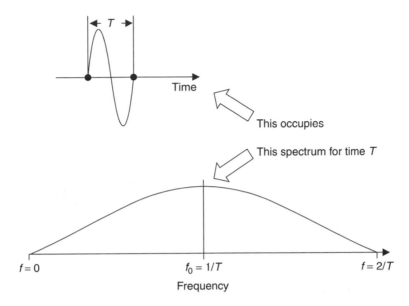

Figure 1.5 The shorter a time signal is, the wider its bandwidth.

signals occupy wide bandwidths, ultrawide bandwidths – but short – and ultrashort slivers of time (see Figure 1.4). The tinier the sliver of time, the wider is the bandwidth of the signal in the radio spectrum as seen in Figure 1.5. When confined to just four cycles of a sine wave, the signal occupies significant bandwidth, on the order of 50%.

Clever coding, modulating, and packing of short signals in *time*, rather than in *frequency*, allows us to separate these desired short signals to distinguish one user from another. This variety of radio signaling is called UWB.

Finally, in Figure 1.6 we see that the entire frequency spectrum can be occupied by multiple users. The users in this case are separated in time rather than in frequency. This is the direct analog of the separation in frequency depicted in Figure 1.3. UWB radio tends to the extreme of separating users in time, while simultaneously occupying large segments of the electromagnetic spectrum.

We see now that there are two ways of sharing the electromagnetic spectrum among many users. The spectrum can be divided in frequency and each user can be assigned a small sliver of the spectrum – a channel. Alternatively, the many users can each occupy the whole spectrum, but for a short sliver of time each. There is, of course, a wide range of intermediate possibilities of separating signals by combinations of time and frequency.

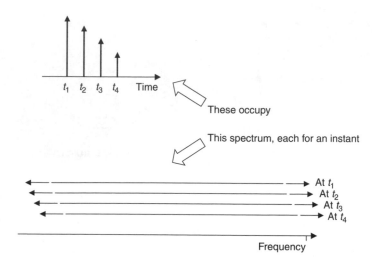

Figure 1.6 Multiple users can each occupy the entire spectrum for a sliver of time.

1.2 The History of Radio

Radio, called wireless at its inception, started out as "ultra-wideband." This could have been entirely by accident or as a consequence of the first transmitters being electromechanical contraptions that generated signals using sparks flying between gaps. From UWB's point of view, the early history of wireless is the story of invention, engineering, and legislation in a relentless march towards narrowband-channelized usage of the radio spectrum.

In 1864, James Clerk Maxwell, while chair at the University of Edinburgh, formulated the concept of electricity and magnetism using the language of mathematics in his equations of electromagnetism. His theory predicted that energy can be transported through materials and through space at a finite velocity by the action of electric and magnetic waves moving through time and space. That finite velocity was, astounding at the time, the velocity of light. The surprise was that this velocity linked light and electromagnetic waves as being the same phenomenon. Maxwell never validated his theory by experiment, and his results were opposed at the time.

Heinrich Rudolf Hertz, starting 22 years later, put into practice what Maxwell proposed with mathematics in a remarkable set of historical experiments [Bryant 1988] spanning the years 1886 to 1891. Hertz calculated that an electric current oscillating in a conducting wire would radiate electromagnetic waves into the surrounding space. In 1886, using a spark-gap apparatus to generate radio energy, he created and detected such oscillations over a distance of several meters in his lab. The radiated waves were dubbed *Hertzian Waves* at the time, and today the basic unit of measuring oscillations per second is the Hertz, abbreviated to Hz. Through these experiments, wireless became a harmonic oscillation game – sine waves. The era of *"wireless"* had begun.

At the turn of the century, the radio arts were developed to practical usability by pioneer inventors and scientists like Alexander S. Popov (also spelled 'Popoff') and Nikola Tesla with their grasp of tuned resonant transmitter and receiver circuits. Popov stated on 7 May 1895 in a lecture before the Russian Physicist Society of St. Petersburg that he had transmitted and received signals across a distance of 600 m. In that same year, Guglielmo Marconi, using a Hertz oscillator, antenna, and receiver very similar to Popov's, successfully transmitted and received signals within the limits of his father's estate at Bologna, Italy. Popov's radio receiving and transmitting system would eventually earn him a Grand Gold Medal for research at the Paris International Exposition of 1900 [Howeth 1963]. On the other hand, the entrepreneur Marconi took his wireless hardware to Britain. In 1896, for his efforts, Marconi was awarded British Patent number 12039 on 2 June in the same year. In 1897, he formed his first company, Wireless Telegraph and Signal Company, in Britain, and began manufacturing wireless sets in 1898. By 1901, Marconi, putting to use his innovations with those of his predecessors, had bridged the 3,000-km distance [Desoto 1936] between St. John's Newfoundland and Cornwall, on the southwest tip of England, using Morse code transmissions of the letter "S." With this achievement, Marconi introduced long-distance communication.

Marconi brought his technology to the United States in 1899 with the Marconi Company. Soon, he controlled patents for the tuner, patented by British inventor Oliver J. Lodge in 1898 [Lodge 1898], and for the John A. Fleming valve (vacuum tube) of 1904 that acted as a diode tube to efficiently detect wireless signals. The Lodge patent is particularly interesting in that it offers advantages in transmitting and receiving "tuning" circuits so that multiple stations may operate side by side in the radio spectrum without mutual interference. The movement was primarily away from wideband signals because at that time there was no way to effectively recover the wideband energy emitted by a spark-gap transmitter. There was also no way to discriminate among many such wideband signals in a receiver. Wideband signals simply caused too much interference with one another to be useful.

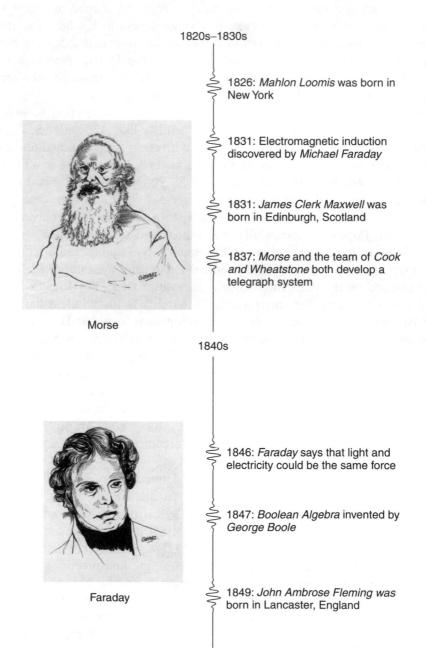

1820s–1830s

1826: *Mahlon Loomis* was born in New York

1831: Electromagnetic induction discovered by *Michael Faraday*

1831: *James Clerk Maxwell* was born in Edinburgh, Scotland

1837: *Morse* and the team of *Cook and Wheatstone* both develop a telegraph system

Morse

1840s

1846: *Faraday* says that light and electricity could be the same force

1847: *Boolean Algebra* invented by *George Boole*

1849: *John Ambrose Fleming* was born in Lancaster, England

Faraday

1850s–1860s

$$\nabla \times \mathbf{E} = -\frac{\partial \mathbf{B}}{\partial t} \qquad \nabla \cdot \mathbf{D} = \rho$$

$$\nabla \times \mathbf{H} = \frac{\partial \mathbf{D}}{\partial t} + \mathbf{J} \qquad \nabla \cdot \mathbf{B} = 0$$

1856: *Nikola Tesla* was born in Croatia

1857: *Heinrich Hertz* born in Hamburg, Germany

1864: *Maxwell* explains the behavior of electromagnetic waves with "Maxwell's Equations"

1865: *Mahlon Loomis* transmits a message between two mountains in Virginia using a wireless telegraph

1866: *Reginald Fessenden* was born in Quebec, Canada

1870s

1872: *Mahlon Loomis* is granted Patent #129, 971 by the US Government for a type of wireless communication

1872: *Fessenden* sends voice approximately one mile using a spark generator

1874: *Guglielmo Marconi* was born in Bologna, Italy

Nikola Tesla

Maxwell

Fessenden

1880s

UWB overview:
—Hertzian experiments
were UWB
—Apparatus was spark gap
—Large RF bandwidths

Hertz

1886–1889: *Hertz* conducted a series
of experiments that proved Maxwell's
theory that light was a form of
electromagnetic radiation

1895: *Alexander S. Popov*
demonstrates radio transmission
using tuned circuits

1900s

UWB overview:
—Wireless is "tuned"

Popov

1901: *Marconi* receives the Morse
code letter "S" in Newfoundland
transmitted by *Fleming* from England

1901: *Fleming* invents the first Tube
known as the "Fleming Valve"

1901: *Heterodyne Reception* is
invented by *Fessenden*

1909: *Karl Braun* and *Marconi* are
jointly awarded the Nobel Prize in
physics for their contribution to
developing wireless telegraphy

1910s–1920s

UWB overview:
—Analog processing
—Long process of innovation

Armstrong

1910: Wireless station installed
in the Eiffel Tower by the French
Army Signal Corps

1912: *Armstrong* invents
regeneration

1912: *US Congress* passes Radio
Act of 1912

1920s

UWB overview:
—Commercial broadcasting
grows

1920: Using a shortwave radio,
Marconi establishes a link between
London and Birmingham, England

1920: *Armstrong* announces the
first superheterodyne circuit

1930s

UWB overview:
—Separation of service by
wavelength

1934: US Congress creates the
*Federal Communications
Commission* (*FCC*) through the
Communications Act

1934: *Armstrong* applies his theory
to FM

1940s

UWB overview:
—Shannon's papers refer
to the "down in the noise"
as most efficient
communication

Claude Shannon

1942: US Patent #2,292,387 issued
to H.K. Markey and George Antheil
for a frequency-hopping technique in
communications

1943: US Supreme Court overturned
Marconi invention of "modern radio"
in favor of Tesla

Hedy Lamarr
(H.K. Markey)

1948: Claude Shannon's *A
Mathematical Theory of
Communication* is published

1950s–1970s

UWB overview:
—1950s UWB & impulse
technology heavily
investigated for
communications, radar &
other applications
—1960s Patents begin
appearing using UWB-
like techniques
—1970s Digital techniques
applied to UWB impulse
radios

1958: *Jack Kilby* produces the first
integrated circuit

1971: *Intel* develops the first
microprocessor

1979: First cellular phone network
begins in Japan

1980s–2000s

UWB overview:
—1980s Publications on
UWB start to appear
—1990s Attempts to make
UWB legal again
—2000s UWB is made legal
in the United States
—International entities are
"on the verge"

1983: The United States starts cellular
phone network

1988: Time Domain Corporation
introduces the FCC to UWB

1993: First UWB chip set created by
Aether Wire & Location, Inc.

UWB overview:
—2002 UWB Masks Defined

1998: FCC notice of inquiry on UWB

1999: FCC waivers for UWB-imaging
systems

2000: Notice of proposed rulemaking
by FCC

2002: FCC approves UWB for
commercialized use

1.3 About the Technology of the Time

Early wireless communications relied on Morse code signaling, which was generated by hand and copied by ear. Morse signaling consists of keying a carrier signal on and off in combinations of dots ("*dits*") and dashes ("*dahs*") that represent alphabetic characters. A moderate messaging rate was about 25 words per minute – which, in today's measure, is equivalent to 20 bps. So the *information bandwidth* of the early wireless signals was relatively small, 10s of Hertz, yet the crude transmitting apparatus emitted very wideband signals, often 100s of kilohertz wide. The consequences were as follows:

1. Signals occupied significantly more spectrum than necessary for communications. Hence, there was significant interference among stations.

2. Receivers were likewise wideband and relatively "deaf" (inefficient). Thus, they collected excess background noise compared to the information bandwidth and could "hear" only the strongest signals. Consequently, the signal-to-noise ratio (SNR) was poor.

With the combination of spectral inefficiency and receiver inefficiency, interference among wireless communicators was a serious issue. Wireless *needed to become narrowband* for survival.

Early Receiver

1.4 Wireless Becomes Radio: The Era of Broadcasting and Regulations

By 1905, Reginald Fessenden of Canada invented a continuous-wave *voice* transmitter using a high-frequency mechanical alternator that was developed by Charles Steinmetz at General Electric in 1903 to generate the radio signal carrier. Fessenden had found a way to change the amplitude of the wireless signal in step with audio amplitude variations: amplitude modulation or AM was born. *Information no longer needed to be broken down into the on/off carrier interruptions according to the Morse*

Code; AM allowed audio to be sent directly on the carrier. He made voice broadcasts from Brant Rock, Massachusetts, on Christmas Eve, 1906, and astonished ship radio operators hundreds of miles out in the Atlantic who heard the *audio* program amid their Morse code *dits* and *dahs*. It would be another 10 years before voice broadcasting became commonplace. It needed inventions and developments like Harold D. Arnold's amplifying vacuum tube in 1913 that made possible coast-to-coast telephony and the first transatlantic radio transmission in 1915.

Prior to 1912, radio was largely the domain of amateur experimenters and ship-to-shore communications for both naval and commercial operations. Interference was a serious problem. Obsolete spark transmitters emitted wideband signals. In the United States, the Radio Act of 23 July 1912 stepped in to mitigate the interference issues but was largely unsuccessful. The Radio Act of 1927 established the Federal Radio Commission (FRC), and the Communications Act of 1934 established the Federal Communications Commission (FCC) giving regulatory powers in both wire-line and radio-based communications. Stations were to be licensed and separated by wavelength, or frequency, and stations were to use a "pure wave" and a "sharp wave" (sine wave carriers) in the words of the FRC. Sine wave communications and narrowband signals were now mandated. Unfiltered spark emissions, dubbed "class B damped sine wave emissions," were prohibited. Radio signals were destined to become "channelized" (see Figure 1.3). These rules *required* that radio signals be narrowband. By organizing the spectrum and controlling interference, the regulations smoothed the way for commercial AM broadcasting to grow.

1.5 Advantages in Wider Bandwidths

The information in AM is encoded by amplitude variations in a carrier. Any other natural amplitude variations, such as amplitude noise, static, and lightning crashes, would add to the desired amplitude modulated information and be perceived as noise and distortion. Edwin Armstrong set himself to the task of finding a way to make broadcast radio insensitive to these amplitude distortions. In 1933, he discovered the advantages of *wideband* frequency modulation (FM) [Armstrong 1933]. In FM modulation, the *frequency* rather than the amplitude of the transmitter carrier was varied in proportion to the *amplitude* of the voice signal. Most importantly, Armstrong realized that an FM signal did not need to have a narrow bandwidth. It could vary over a wide range, several times as wide as an AM signal, and as a result have a far better SNR than AM. This meant that programs broadcasted using wideband FM could be made higher fidelity and less distorted than AM broadcasts.

Armstrong's discovery laid the foundations for information theory, which quantifies how signal bandwidth can be exchanged for noise immunity, that is, for a reduction in amplitude noise distortion. *Voice and music transmissions could now be static free.* The intentional and controlled bandwidth spreading of a signal beyond its information bandwidth was shown to have significant desirable benefits – this was a small but very important challenge to the narrowband mantra.

Commercial broadcast interests developed along channelized services in the AM broadcast band, and, later, in wider channel bandwidths in the FM broadcast band. Specific allocations in the frequency spectrum were established for radio amateurs, for broadcasting, and, later on, for television and personal communications. Wireless, now radio, communications were becoming a practical reality. The radio frequencies of interest to personal communications were steadily evolving into voice communications using analog modulations: AM and FM, both narrow and wideband. By the mid-1930s, the era of two-way *radio* communications in the low VHF range (30 to 40 MHz) became a reality. FM, developed by Edwin Armstrong and championed by Dan Noble for two-way land-mobile communications, effectively opened up the VHF bands for economical communications systems. By the mid-1940s, radio frequencies for land-mobile communications were allocated in the 150-MHz range. This was followed by the allocation of frequencies in the 450-MHz range during the decade of the 1960s. As the pressure increased for more and more radio signaling and radio services, higher and higher frequencies in the radio spectrum were being assigned, channelized, and developed.

1.6 Radio Takes Another Wider-band Step

Traditionally, the FCC had favored narrowband radios, which concentrate all of their power in fairly narrow channels within the radio frequency spectrum. However, as the number of users sharing the spectrum was increased, the number of available channels became limited. Claude Shannon, in 1948, offered a new paradigm, redefining the relationship among power density, noise, and information capacity [Shannon 1948].

Shannon said that under certain specific conditions, the more an information signal is spread in bandwidth in a way that makes the signal resemble background noise, the more information it is capable of holding. Because one signal spread in this way resembles noise to another signal that is similarly spread, both can coexist because, under some specific conditions, signal energy can be detected more efficiently than noise energy. Thus, with wider bandwidth, more such sharing can occur and more total information can be conveyed. Hence, an alternative to transmitting a signal with a high power density and low bandwidth would be to use a low power density and a wide bandwidth [Malik 2001].

Shannon's observations led to "spread-spectrum" modulation in which the signals are intentionally spread using a special family of digital codes to many times their information bandwidth. Special digital codes are used to distinguish multiple users that are simultaneously sharing the same band. Spread-spectrum technology applied to cellular telephone system resulted in a change in spectrum-regulation policy. It was the *second* time in the history of radio that the advantages of wideband signaling was recognized as important enough to result in changes in the regulations away from the narrowband mantra. This time, the FCC had allocated a *block* of spectrum within which multiple users shared the entire block of spectrum by overlapping signals across the entire band, rather than by allocating narrow slivers of bandwidth per user. Spread-spectrum users would be separated by coding rather than by frequency channels. Because of the increased efficiency in the use of precious (and expensive) spectrum, this led to improvements in the capacity of cellular systems that in turn reduced the cost of spread-spectrum cellular services.

Today, a significant growth in personal communications is taking place using spread-spectrum techniques in blocks of spectrum that have been set aside for unlicensed operations shared by other users. These signals appear covert and coexist well with other signals transmitted in the same frequency bands. This method makes much more efficient use of the congested spectra and allows greatly expanded utilization. The modern era of "*digital wireless*" has begun.

1.7 Still Wider has More Advantages

Through the years, a small cadre of scientists has worked to develop various techniques of sending and receiving short-impulse signals between antennas. Impulses are short time signals – the shorter the impulse, the wider its bandwidth. The experiments led to "impulse radio," later dubbed UWB radio. By the late 1960s and 1970s ([Harmuth 1968] and [Harmuth 1978]), the virtues of wideband nonsinusoidal communications were being investigated for nongovernment uses. Prior to that, the primary focus was on impulse radar techniques and government sponsored projects. In the late 1970s and 1980s, the practicality of modern low-power impulse radio techniques for communications and positioning/location was demonstrated using a time-coded time-modulated approach by Fullerton [Fullerton 1989], and later by others [Fleming 1998] using UWB spread-spectrum impulse techniques. Digital impulse radio, the modern echo of the Hertz and Marconi century-old spark transmissions, now reemerges as ultra-wideband radio. On 14 February 2002 [FCC15 2002], the FCC adopted the formal rule changes officially permitting ultra-wideband operations. The ruling defines access to a 7,500-MHz-wide swath of unlicensed spectrum between 3.1 and 10.6 GHz that is made available for commercial communications development in the United States.

1.8 Summary

Wireless began as wideband signaling – UWB by today's measure and definition – because the transmitters of the time were spark-gap devices that emitted wideband, noisy signals. The receivers in use at the time were simple amplitude detectors that could not efficiently gather the wideband energy. This resulted in an inferior SNR performance, hence requiring large transmitter powers to achieve desired ranges. High-transmitter powers and excessively wide signal bandwidths meant significant spectrum sharing problems and plenty of interference. The receiver efficiency and interference issues drove wireless to narrower and narrower bandwidths per signal. The ideal was a signal as narrow as the information bandwidth. Regulations in 1912 mandated the narrowest bandwidths possible, and codified the separation of wireless services by wavelength (frequency).

In 1933, the advantages of intentional and controlled signal widening to many times the information bandwidth were discovered in the form of wideband FM radio. In this approach, bandwidth could be exchanged for noise immunity – to the delight of the FM broadcast industry. Since 1912, all spectra were allocated on a per channel basis per user to the exclusion of other users, and emissions were to be kept to the narrowest practical bandwidth. Then came spread-spectrum technology. By 1985, the FCC began allowing spectrum technology in which multiple users would be separated by direct-sequence codes rather than by discrete frequency channels. Commercial deployment of Code Division Multiple Access (CDMA) spread-spectrum cellular telephony followed in 1995. In 1999, the International Telecommunication Union adopted an industry standard for third-generation (3G) wireless systems that can deliver high-speed data and other new features. The 3G standard includes three operating modes based on CDMA technology. Thus, spectrum policy shifted away from "separation by wavelength." However, those multiple users were in a block of spectrum allocated for that purpose.

Throughout the last half of the twentieth century, much experimentation and development took place in wideband impulse radar transmissions – the forerunner of modern UWB. Independently, commercial experiments, inventions and petitions before the FCC in the 1980s and 1990s led to the landmark FCC regulations of 2002 to permit low-power UWB for commercial development. This was a major shift in spectrum policy. Under the new regulations, multiple unlicensed users could share spectrum previously allocated to other users, including licensed users, on a noninterference basis. Thus, UWB is as much about an exciting new technology as it is about the unprecedented, unlicensed access to a huge amount of spectrum. Standards are being developed, and the UWB industry is on the verge of market deployment.

References

[Armstrong 1933] E. H. Armstrong, *Radio Signaling System*, U.S. Patent 1,941,066, 26 December 1933.

[Bryant 1988] J. H. Bryant, *Heinrich Hertz – The Beginning of Microwaves*, New York: IEEE Press, 1988.

[DeSoto 1936] C. B. DeSoto, *Two Hundred Meters and Down*, Newington, CN: The American Radio Relay League, 1936.

[FCC15 2002] US 47 CFR Part15 Ultra-Wideband Operations FCC Report and Order, 22 April 2002.

[Fleming 1998] R. A. Fleming and C. E. Kushner, *Spread Spectrum Localizers*, U. S. Patent 5,748,891, 5 May 1998.

[Fullerton 1989] L. Fullerton, *Time Domain Transmission System*, U.S. Patent 4,813,057, 14 March 1989.

[Harmuth 1968] H. F. Harmuth, "A generalized concept of frequency and some applications", *IEEE Transactions on Information Theory*, **IT-14**(3), 375–381, 1968.

[Harmuth 1978] H. F. Harmuth, "Frequency-sharing and spread-spectrum transmission with large relative bandwidth", *IEEE Transactions on Electromagnetic Compatibility*, **EMC-20**(1), 232–239, 1978.

[Howeth 1963] L. S. Howeth, *History of Communications-Electronics in the United States Navy*, Washington, DC: Bureau of Ships and Naval History, 1963, (Online): <http://earlyradiohistory.us/1963hw.htm>.

[Lodge 1898] O. J. Lodge, *Electric Telegraph*, U.S. Patent 609,154, 16 August 1898.

[Malik 2001] R. Malik, *Spread Spectrum, Secret Military Technology to 3G*, (Online): <http://www.ieee.org/organizations/history_center/ cht_papers/SpreadSpectrum.pdf> 7 August 2001.

[McKeown 2003] D. McKeown, *Gammz UWB Cartoons and Art*, Private Communication to K. Siwiak, December 2003.

[Shannon 1948] C. E. Shannon, "A mathematical theory of communication", *The Bell System Technical Journal*, **27**, 379–423, 623–656, 1948.

Further Reading

[FRBH 2003] *History of UWB Technology*, (Online): <http://www. multispectral.com/history.html>, 1 June 2003.

[Harrington 1961] R. F. Harrington, *Time Harmonic Electromagnetic Fields*, New York: McGraw- Hill, 1961.

[IEEE145 1993] *IEEE Standard Definitions of Terms for Antennas*, IEEE Std 145–1993, SH16279, 18 March 1993.

[IEEE211 1997] *IEEE Standard Definitions of Terms for Radio Wave Propagation*, IEEE Std 211–1997, 9 December 1997.

[Russia 2003] The Russian UWB Group, (Online): <http://www.uwbgroup. ru/eng/common/uwb.htm>, 1 June 2003.

2

The Regulatory Climate

Introduction

For more than a century, and around the globe, the interplay between the evolution of the radio arts and regulations have shaped the way users share the precious and limited radio spectrum. Regulations have provided the mediation that frames how the many users of the electromagnetic spectrum can coexist. The focus of the earliest regulations defined rules that, in step with the way the radio arts developed, made radio signals as narrowband as possible. Inefficient wideband signals were contrary to the progress possible at that time. Modern advancements in the generation, transmission, and reception of wideband signals, however, have improved the manner in which we can efficiently utilize the spectrum. Regulations again, in step with these modern developments, have adapted to permit modern UWB signaling. Regulations address the issues of separation and coordination of and interference among spectrum users. They define the rules for accessing the radio spectrum. The traditional allocation of spectrum had not anticipated the modern development of UWB. However, the arrival of the technology has reshaped the concepts of spectrum management to allow modern UWB technology.

2.1 Electromagnetic Spectrum: "Separation by Wavelength"

The electromagnetic or radio spectrum is an orderly arrangement of radio frequencies or channels arranged like colors in the rainbow. Each segment in this spectrum (see Table 2.1) has natural physical characteristics.

Ultra-Wideband Radio Technology   Kazimierz Siwiak and Debra McKeown
© 2004 John Wiley & Sons, Ltd   ISBN: 0-470-85931-8

Table 2.1 The frequency spectrum and some of its uses.

Frequencies	Band	Characteristics	Uses
3–30 kHz	ELF, VLF	High atmospheric noise, earth–ionosphere wave guide modes, antennas very inefficient	Submarine, navigation, sonar, long-range navigation
30–300 kHz	LF	High atmospheric noise, earth–ionosphere wave guide modes, absorption in the ionosphere	Long-range navigational beacons
0.3–3 MHz	MF	High atmospheric noise, good ground-wave propagation, earth's magnetic field cyclotron noise	Navigation, maritime communication, AM broadcasting
3–30 MHz	HF	Moderate atmospheric noise, ionosphere reflections provide long-distance links. Affected by solar flux density	International shortwave broadcasting, ship to shore, telephone, amateur radio
30–300 MHz	VHF	Meteor scatter possible, normal propagation basically line of sight	Mobile, TV, FM broadcasting, aviation, radio navigation aids
0.3–3 GHz	UHF	Line-of-sight propagation	TV, radar, mobile phone/radio, satellite links, WLANs
3–30 GHz	SHF	Line-of-sight propagation, atmospheric absorption at upper frequencies	Radar, microwave links, satellite links, WLANs, UWB
30–300 GHz	EHF	Line-of-sight propagation, high atmospheric absorption	Radar, military communication, satellite links
300–10^7 GHz	IR-optics	Line-of-sight propagation, high atmospheric absorption	Optical communications

Source: After K. Siwiak, *Radiowave Propagation and Antennas for Personal Communications*, Second Edition, Norwood, MA: Artech House, 1998.

The lowest frequencies, longest wavelengths, extremely low frequencies (ELF), and very low frequencies (VLF) are characterized by propagation that fills and is contained within two concentric spherical shells. One shell is the surface of the earth and the other is the ionosphere of ionized atmospheric gases some 100 km above the earth surface.

ELF and VLF are very susceptible to lightning and atmospheric noises. There is also limited bandwidth as the range extends from a few kilohertz to 10s of kilohertz. The low frequencies (LF) and medium frequencies (MF) extend from 10s of kilohertz to a few megahertz. Radio grew and matured in the LF and MF spectrum because

1. the frequencies are low enough that relatively simple transmitters and receivers can be built, and

2. radio wave propagation, although very much affected by atmospheric noise, is still very suitable for navigation beacons and commercial broadcasting.

Frequencies in the LF and MF ranges are absorbed by the earth's ionosphere, effectively limiting the communication range from 10s to 100s of kilometers relying on ground-wave propagation. In the early history of radio, MF spectrum was thought to be the upper limit of useful radio frequencies. Ground-wave propagation relies strongly on radio currents induced in the earth ground and becomes less effective at higher frequencies because the wavelengths are shorter. Higher frequencies than the MF range were left to experimenters and radio amateurs by regulatory agencies around the world. Signals in the high frequencies (HF) range from a few megahertz to tens of megahertz are capable of reflecting from earth's atmosphere. Long-distance and transcontinental communications became possible in this band when experimenters discovered ionospheric reflections. International shortwave broadcasting continues to this day in the HF range. In the upper reaches of the HF band, however, the ionosphere becomes transparent to radio signals.

Radio components and construction techniques are more difficult and critical as frequencies are increased. The use of very high frequencies (VHF) and ultra high frequencies (UHF) began to mature as

1. the technology improved,

2. the lower frequencies became overcrowded.

Signals at VHF and UHF penetrate the ionosphere, so they are suitable for earth–space links as well as for local communications. Because there is comparatively more bandwidth in these upper decades of the spectrum, many bandwidth hungry commercial broadcasting services are found in this range. For example, FM broadcasting and television signals appear between 54 and 806 MHz. Commercial services in the frequency ranges above a few gigahertz or so have also recently begun to appear. These are also the frequencies used by the most interesting wireless data devices. The UWB spectrum that is useful for communications and measurement systems begins at 3.1 GHz and ends at 10.6 GHz in the Super High Frequencies (SHF).

The electromagnetic spectrum begins at the very lowest frequencies, at several Hertz, where the earth–ionosphere concentric shell resonates, and continues through the radio frequencies, microwaves, and visible light. As pictured in Figure 1.2 of the previous chapter, it continues through the X-ray frequencies, on up to where waves are indistinguishable from elementary atomic particles. The segment of the electromagnetic spectrum that is usually regulated by various national agencies starts in the vicinity of 3 kHz and extends to approximately 300 GHz. The commercial deployment of UWB communication devices appears in the 3.1- to 10.6-GHz band, well within traditionally regulated spectrum.

2.2 Radio Regulations

Early radio devices were developed from simple electromechanical components. Their signals were sloppy and used spectrum wastefully. As usage and the user population increased, interference among radio operators precipitated two things:

1. Spectrum usage at higher and higher frequencies

2. The need to regulate spectrum usage and the quality of radio transmissions.

Regulations began to separate users by wavelength or operating frequency. Also, to conserve the precious spectrum, users were obligated to use the narrowest possible bandwidth. The technology for the "narrowest bandwidth" was thus encouraged by regulations to grow and mature. Subsequent newer technologies, such as wideband FM, needed to show an exceptional superiority in some aspect of operation to be acceptable to regulatory bodies. For wideband FM, that superior advantage was a significantly improved quality of audio broadcasting. For narrower band FM two-way radio communications systems, the advantage was in the low commercial cost of implementation at VHF and UHF compared with the much narrower band single-sideband (SSB) approaches. The technology, regulation, and channelization of spectrum all matured along a path of narrowband radio set in the early 1900s.

Prior to 1912, radio was largely the domain of amateur experimenters and ship-to-shore communications for both naval and commercial operations. Interference was a serious problem. Obsolete spark transmitters emitted very wideband signals. The Radio Act of 23 July 1912 stepped in to mitigate the interference issues, and required the registration of transmitters with the Department of Commerce but did not provide for the control of their frequencies, operating times, or of the station transmitter power. Thus, there was no real regulatory power and the 1912 Act was largely unsuccessful; however, the march to narrowband had begun.

The March to Narrowband

In 1922, US government users of the spectrum, under the Secretary of Commerce formed the Inter-department Radio Advisory Committee (IRAC) to coordinate government use of the spectrum [Roosa 1992]. The Radio Act of 1927 established the Federal Radio Commission (FRC), and the Communications Act of 1934 established the Federal Communications Commission (FCC). The 1934 Act gave the FCC broad regulatory powers in both wire-line-based communications, such as telephone and telegraph systems and radio-based communications, limited at the time to broadcasting, long-distance single-channel voice communications, maritime and aeronautical communications, and experiments that led to radar and television applications. Section 305 of the Act preserves for the President the authority to assign frequencies to all Federal Government owned or operated radio stations. In addition, the President retains the authority to assign frequencies to foreign embassies in the Washington, D.C., area and to regulate the characteristics and permissible uses of the Government's radio equipment. The IRAC, whose existence and actions were affirmed by the President in 1927, has continued to advise those exercising responsibility for the Section 305 powers of the

President. These powers currently are delegated to the Assistant Secretary of Commerce for Communications and Information who is also the Administrator of the National Telecommunications and Information Administration (NTIA).

The use of the electromagnetic spectrum in the United States is managed using a dual organizational structure (see Figure 2.1). NTIA manages the Federal Government's use of the spectrum while the FCC manages all other uses. The Act provides for developing classes of radio service, allocating frequency bands to the various services, and authorizing frequency use. However, the Act does not mandate specific allocations of bands for exclusive Federal or non-Federal use. All such allocations stem from agreements between NTIA and the FCC. In other words, there are no statutory "Federal" or "non-Federal" bands. Because UWB emissions would span across both Federal and non-Federal users, the regulations governing UWB needed coordination and agreement between the NTIA and FCC.

Regulations have turned the spectrum into an orderly arrangement giving various radio services their own frequency ranges. Many of these services are coordinated internationally because radio signals respect no political boundaries and wandering signals can create interference. Since 1912, spectrum users traditionally have been separated from each other on the basis of wavelength, or frequency. Since perfect separation is not possible, regulations also set the technical standards addressing the level of

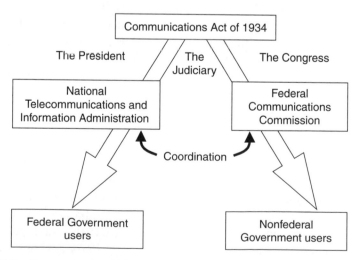

Figure 2.1 Spectrum regulation in the United States is managed by both the FCC and the NTIA.

interference tolerable among users. Some services, such as safety-of-life radio services, are more sensitive and more important than others. They are accordingly afforded more protection. Regulations also address the levels of unintentional noise, or unintentional radio emissions that are caused by electrical and electronic equipment. It is in the context of the levels permitted for the unintentional emitters that regulations permitting modern UWB were developed.

2.3 Adoption of UWB in the United States

Research in communications and in radio technology began to reveal that the radio frequency spectrum could be shared among many users more effectively than by allocating fixed narrowband channels. Consequently, regulations have appeared since 1985, which assigned blocks of nonchannelized spectrum to mobile service providers. The mobile telephone service provider satisfies the demands of many simultaneous users by spreading each of their signals over the entire allocated bandwidth. Everybody utilizes the entire spectrum, and the separation of users is accomplished by using special digital coding and modulation techniques. Simultaneous usage of this dedicated block of spectrum by multiple users facilitates spectral efficiency. Starting with petitions in the 1980s and culminating in early 2002, regulations in the United States were adopted that permitted the use of ultra-wideband signals within a huge block of spectrum between 3.1 and 10.6 GHz at power levels commensurate with those of unintentional emitters. The commercial deployment of UWB communications, ranging and imaging devices are permitted in the United States on a noninterference basis over the band already occupied by existing radio services.

Jurisdiction over the radio frequency spectrum in the United States is shared by the FCC, an Executive Branch agency reporting to The President, and the NTIA, a Department of Commerce reporting to The Congress. The FCC oversees the non-Federal users, and allocates spectrum to commercial, private, amateur, state, and local public safety departments. The NTIA oversees spectrum for Federal government users. The two agencies follow a *Memorandum of Understanding on Joint Use Spectrum*. A reciprocal agreement provides that the NTIA and FCC will give

"notice of all proposed actions which would tend to cause interference to Non-Government and Government station operations. [This] notification must be given in time for [NTIA/FCC] to comment prior to final action...Final action by either agency will not, however, require approval by the other agency."

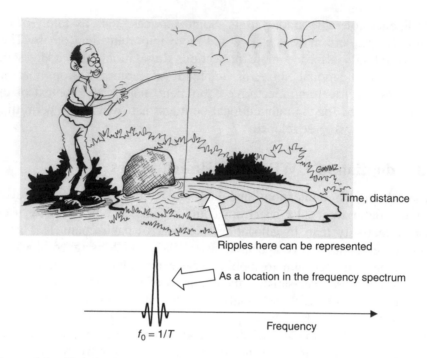

Figure 2.2 Signals like ripples in a pond are seen by intended and unintended eyes [McKeown 2003].

Thus, neither the NTIA nor the FCC, who appear in different branches of the U.S. government, can act unilaterally in spectrum regulatory matters.

Signals, once emitted, spread like ripples in a pond from a source to the intended receiver, and in all other directions (see Figure 2.2). Radio signals do not respect the boundaries of just a desired link. A transmission intended for one receiver, once sent, continues to expand and fill our universe. Radio signals are available to the intended user, but like ripples in a pond reaching all shores also invade the space of every other user who might view that signal as interference. To others, the signal is also available as a target for eavesdropping. These signals respect no international boundaries either. These issues of privacy, interference, and sharing of the spectrum resource highlight the need for agreements among users, and for regulations that referee and manage access to that resource. The FCC mandate from the Code of Federal Regulations (47 USC 157 – *New technologies and services*) states:

(a) It shall be the policy of the United States to encourage the provision of new technologies and services to the public. Any person or party (other than the Commission) who opposes a new technology or service proposed to be permitted under this chapter shall have the burden to demonstrate that such proposal is inconsistent with the public interest.

(b) The Commission shall determine whether any new technology or service proposed in a petition or application is in the public interest within one year after such petition or application is filed. If the Commission initiates its own proceeding for a new technology or service, such proceeding shall be completed within 12 months after it is initiated.

UWB devices emit energy across wide swaths of spectrum used by incumbent services. They cannot avoid emitting energy into both Government and non-Government spectrum; thus, the FCC must coordinate rules with NTIA. The NTIA is principally concerned with emissions into "restricted" frequency bands that include national security and safety-of-life operations. It is in this climate that a Notice of Inquiry was begun in 1998 by the FCC. Waivers were granted in June of 1999 for three UWB devices. They were as follows:

1. Time Domain for through-wall imaging device

2. Zircon for a "stud-finder" for rebar in concrete

3. US Radar for a ground-penetrating radar.

A Notice of Proposed Rule Making was adopted in May 2000, which culminated in a First Report and Order (R&O) adopted on 14 February 2002 and released (with slight modifications) on 22 April 2002 (see Appendix A and [FCC15 2002]). The public inquiry, which had begun in 1998, drew over 1,000 documents filed, and the NTIA commissioned several reports that reviewed interference potential to Government systems. The FCC objectives for the UWB Report and Order were to enable the introduction of UWB technology while protecting incumbent users against harmful interference. Thus, the intent was to provide numerous benefits to the public and to maintain US technical leadership, while establishing appropriate interference standards.

2.4 Summary of First Report and Order

The FCC ruling provides significant protection for sensitive systems like the Global Positioning Service (GPS), aviation systems, and safety-of-life services. It incorporates NTIA recommendations and allows UWB technology to coexist with existing radio services without causing harmful interference. The R&O establishes different technical standards and operating restrictions for three types of UWB devices based on their potential to cause interference. These three types of UWB devices are

1. imaging systems including Ground Penetrating Radars (GPRs) and through-wall radar, surveillance, and medical imaging devices,

2. vehicular radar systems,

3. communications and measurement systems.

Appendix A contains the text of the FCC Report and Order. The *communications and measurement systems* category is of primary interest to commercial UWB radio technology, and is discussed further here. UWB operation is defined as a transmission system having UWB transmitter and is defined (see Appendix A for details) in terms of the following:

(a) UWB Bandwidth: the frequency band bounded by the points that are 10 dB below the highest radiated emission, as based on the complete transmission system including the antenna. The upper boundary is designated f_H and the lower boundary is designated f_L. The frequency at which the highest radiated emission occurs is designated f_M.

(b) Center frequency: $f_C = (f_H + f_L)/2$.

(c) Fractional bandwidth: $BW = 2(f_H - f_L)/(f_H + f_L)$.

(d) UWB transmitter is an intentional radiator that, at any point in time, has a fractional bandwidth $BW \geq 0.20$ or has a UWB bandwidth $\geq$ 500 MHz, regardless of the fractional bandwidth.

(e) EIRP: Equivalent isotropically radiated power, that is, the product of the power supplied to the antenna and the antenna gain in a given direction relative to an isotropic antenna.

(f) Handheld device is a portable device, such as a laptop computer or a PDA, that is primarily handheld while being operated and that does not employ a fixed infrastructure.

The regulations permit indoor UWB systems and handheld devices. The emission limits for handheld devices are shown in Figure 2.3, and are compared to indoor limits in Table 2.2.

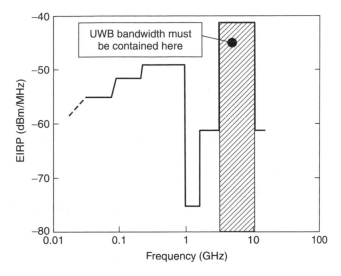

Figure 2.3 The FCC emission masks for handheld UWB devices.

Table 2.2 FCC limits for indoor and handheld systems.

Frequency (MHz)	Indoor EIRP (dBm)	Handheld EIRP (dBm)
Below 960	Section 15.209	Section 15.209
960–1,610	−75.3	−75.3
1,610–1,990	−53.3	−63.3
990–3,100	−51.3	−61.3
3,100–10,600	−41.3	−41.3
Above 10,600	−51.3	−61.3

Indoor UWB devices, by the nature of their design, must be capable of operation only indoors. The emissions from indoor devices shall not be intentionally directed outside of the building in which the equipment is located, such as through a window or a doorway, to perform an outside function, such as the detection of persons about to enter a building. A communications system shall transmit only when the intentional radiator is sending information to an associated receiver. The UWB bandwidth for indoor devices must be contained between 3,100 and 10,600 MHz. The radiated emissions above 960 MHz from a device operating under the provisions of this section shall not exceed the average limits shown in Table 2.1 when measured using a resolution bandwidth of 1 MHz.

UWB handheld devices are relatively small devices that do not employ a fixed infrastructure. These devices shall transmit only when sending information to an associated receiver. Antennas may not be mounted on outdoor structures such as the outside of a building or on a telephone pole. Antennas may be mounted only on the handheld UWB device. Handheld UWB devices may operate indoors or outdoors. The UWB bandwidth of a handheld device must be contained between 3,100 and 10,600 MHz. Additional restrictions and details of indoor and handheld UWB devices are contained in Appendix A.

Even as the FCC issued the Report and Order, the FCC commissioners stated, "We are concerned, however, that the standards we are adopting may be overprotective and could unnecessarily constrain the development of UWB technology. Accordingly, within the next 6 to 12 months we intend to review the standards for UWB devices and issue a further rule making to explore more flexible technical standards and to address the operation of additional types of UWB operations and technology." The FCC has thus drawn a delicate balance between enabling the introduction of a new technology with huge market potential, and respecting the needs of incumbent spectrum users.

2.5 Regulations in Asia: The UFZ in Singapore

Researchers in Japan, Korea, China, Singapore, Taiwan, and perhaps other countries are showing an interest in UWB radio technology by their active participation in IEEE standards activities relevant to UWB technology. Currently in Asia only the Infocomm Development Authority (IDA) in Singapore permits UWB, and then only with a special experimental license. The IDA recognizes the market potential of UWB and encourages Singapore-based companies to collaborate with key global technology providers to engage in UWB research and development, product design, manufacturing, and user trials. Non-Singaporean UWB players are welcome to establish a presence in Singapore, to use Singapore as a test bed for their cutting-edge technologies and as a springboard to the regional UWB market. According to Ching-Yee Tan, CEO, IDA of Singapore (see [IDA 2003]),

> "Ultra-Wideband (UWB) is a highly promising but potentially disruptive wireless technology. Highly promising, because of its wide range of applications, but also disruptive because it could require a new way to allocate and

use radio frequency spectrum. Launched on 25 February 2003, IDA's UWB program is a two-year focused effort to bring UWB technology to Singapore with the following three key thrusts:

First: IDA will issue trial permits for controlled UWB emissions within a specific area known as the UWB Friendly Zone (UFZ), giving developers substantial latitude in experimenting with newer and more innovative UWB designs.

Second: IDA will build upon the existing pool of UWB knowledge by conducting a series of UWB compatibility studies. These findings will form the basis of subsequent regulation to permit the commercial use of UWB in Singapore.

Third: IDA will promote the growth of a vibrant UWB ecosystem comprising holders of UWB intellectual property rights, technology providers, IC design houses, semiconductor foundries, consumer product manufacturers, venture capitalists, and a community of early adopters."

The UWB Friendly Zone (UFZ) is located within Science Park II, amidst the research, development, and engineering community in Singapore. UWB technology providers and their partners are able to establish facilities in the UFZ and work closely with local researchers.

In order to facilitate experimentation and encourage innovation, IDA issues trial licenses to companies permitting them to operate UWB devices within the UFZ, subject to the emission limits given in Table 2.3.

The emissions of Table 2.3 are measured in a resolution bandwidth of 1 MHz using an RMS detector with a video integration time of 1 ms or less. The limits are 6 dB less stringent, and have an expanded lower frequency band edge than what is permitted by the FCC. Figure 2.4 shows the UFZ limits in comparison to the FCC emissions mask.

Table 2.3 UWB limits for the Singapore UFZ.

Frequency (MHz)	EIRP (dBm)
Below 960	Not intentional
960–1,610	−75.3
1,610–1,990	−63.3
1,990–2,200	−61.3
2,200–10,600	−35.3
Above 10,600	−41.3

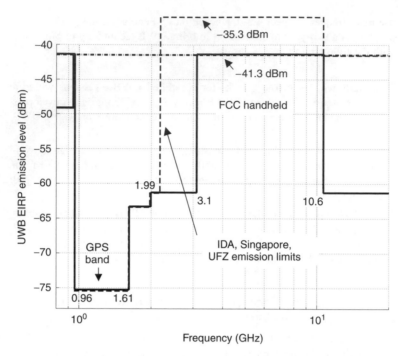

Figure 2.4 The Singapore UFZ emission limits compared with FCC handheld limits.

2.6 Regulation Activities in the European Union (EU)

Within the European Union (EU), the European Radiocommunications Office (ERO) is the facilitator for the European Technical Standards Institute (ETSI) and for the European Conference of Postal and Telecommunications Administration (CEPT), which deals with UWB spectrum sharing and compatibility studies (see Figure 2.5). The groups within CEPT with prime responsibility are as follows:

1. Short Range Devices (SRD) Management Group, who are mandated to develop a draft of UWB regulation;

2. Spectrum Engineering SE 24, mandated to perform spectrum sharing studies especially focused on SRD UWB concepts;

3. Spectrum Engineering SE 21, mandated to perform spectrum studies focused on UWB unwanted emissions;

4. Spectrum Engineering (SE) Working Group (WG), who are to approve studies of Project Teams 24 and 21.

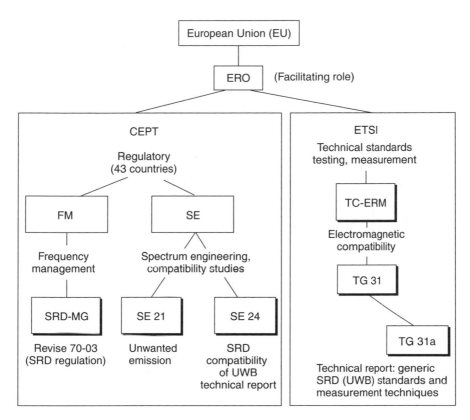

Figure 2.5   European organizations involved in regulation and standards.

ETSI (see [ETSI 2003]), deals with technical standards and electromagnetic compatibility issues. The attitude is very conservative and proposed draft emission standards are significantly more restrictive than the U.S. FCC regulations. Figure 2.6 shows the proposed ETSI emission limits for indoor and for handheld systems. The proposed ETSI indoor limit in the UWB band is the same as the FCC handheld limit, while the ETSI handheld proposal is as much as 20 dB more restrictive.

2.7 Summary

Regulations respond to technology and market needs. The earliest regulations responded to, and drove technology along a narrowband path, because it was the right thing to do at that time. When wider bandwidths for certain radio signaling methods were found to be superior to other methods, those technologies were codified in new regulations. With

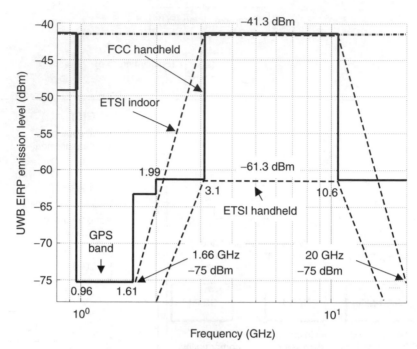

Figure 2.6 Proposed ETSI emission standards for handheld and indoor UWB systems.

UWB radio technology, the *concept* of frequency management itself has been challenged, and regulators, with their supporting organizations, are responding to meet this challenge. The reason that the regulation changes adopting UWB happened more quickly in the United States than in other places is that US companies have devices they want to get to the market place, and that a few key commercial players really pushed the legislation. In particular, Time Domain Corporation, followed by many other interested parties, initially brought the matter to the attention of the FCC. The shear weight of commercial interest in the technology caused the FCC to respond by enabling UWB with a balanced and fair set of regulations appropriate for the United States. Regulating bodies respond to market pressure and other countries will develop their own regulations in their own fashion, appropriate to their own market demands. UWB, though an old technology, is in its modern infancy so there is no doubt that other regulations will spring up all over to accommodate UWB's growth and changes through future innovations and consumer demand.

References

[ETSI 2003] ETSI, *Harmonised Standards Covering Ultrawide Band (UWB) Applications*, Directorate General of the European Commission, Standardisation Mandate: DG ENTR/G/3M/329, Brussels, 25 February 2003.

[FCC15 2002] US 47 CFR Part15 Ultra-Wideband Operations FCC Report and Order, 22 April 2002.

[IDA 2003] IDA (Infocomm Development Authority of Singapore), *Singapore Ultra-Wideband Programme*, Singapore 038988.

[McKeown 2003] D. McKeown, *Gammz UWB Cartoons and Art*, Private Communication to K. Siwiak, December 2003.

[Roosa 1992] P. C. Roosa Jr., *Federal Spectrum Management: A Guide to the NTIA Process*, NTIA Special Publication 91–25, August 1992, (Online): <http://www.ntia.doc.gov/osmhome/primer.html>.

[Siwiak 1998] K. Siwiak, *Radiowave Propagation and Antennas for Personal Communications*, Second Edition, Norwood, MA: Artech House, 1998.

3

UWB in Standards

Introduction

A standard is an agreed-upon definition or format that has been approved
by a recognized organization or is accepted as a norm by the industry. Standards may be *de jure* or *de facto*, that is, they may be set by
official standards organizations, or they may be unofficial standards that
are established by common use. Conformance to standards makes it possible for different manufacturers to create products that are compatible
or interchangeable with each other. Standards make possible the wide
acceptance and dissemination of products from multiple manufacturers
and vendors with an economy of scale that reduces costs to consumers.
Several well-known standards organizations are as follows:

  ANSI (American National Standards Institute)

  CCIR (International Radio Consultative Committee)

  ETSI (European Technical Standards Institute)

  IEEE (Institute of Electrical and Electronic Engineers)

  ISO (International Standards Organization)

  ITU (International Telecommunication Union).

These bodies have produced many easily recognizable standards. For
example, ANSI has defined standards for programming languages such
as C, COBOL, and FORTRAN. CCIR is a body appointed by the United
Nations to recommend worldwide standards in radio matters. Their standards include Radiopaging Code No. 1 (RPC1), which is more widely
recognized as POCSAG. ETSI have produced the GSM (Global System

Ultra-Wideband Radio Technology   Kazimierz Siwiak and Debra McKeown
© 2004 John Wiley & Sons, Ltd   ISBN: 0-470-85931-8

for Mobile Communication) digital cellular telephone standards, which is in use globally. IEEE maintains approximately 800 standards, including IEEE Standards for Local and Metropolitan Area Networks encompass the 802.11 and 802.15 families of communications standards. ISO has defined the ISO 9000 and ISO 14000 families of standards, which are implemented by some 610,000 organizations in 160 countries. ISO 9000 deals with quality management. The ISO 14000 family is primarily concerned with environmental management. ITU defines international standards, particularly communications protocols such as the V.22, V.32, V.34, and V.42, which are protocols for transmitting data over telephone lines.

In addition to the *de jure* standards (approved by organizations), there are also *de facto* standards. These are formats that have become standard simply because a large number of companies have agreed to use them. They have not been formally approved as standards, but they are standards nonetheless. Adobe's PostScript printer-control language is a good example of a *de facto* standard. Another example is Motorola's FLEX paging protocol, which became the worldwide *de facto* standard for high-speed paging.

In UWB matters, the IEEE is active in defining a UWB radio physical layer standard. ETSI is grappling with UWB conformance testing standards in support of European regulatory action. There is wide global involvement in the IEEE standards process, which is likely to influence regulations and perhaps standards around the world.

3.1 High Data Rate UWB Standards Activities in IEEE

A UWB radio physical layer standard is currently under development within the IEEE 802 LAN/MAN Standards Committee. This committee is concerned with Local Area Network (LAN) standards and Metropolitan Area Network (MAN) standards. The most widely used standards are for the Ethernet family, Token Ring, Wireless LAN, Bridging, and Virtual Bridged LANs. The IEEE 802.15.3a Wireless Personal Area Networks (W-PAN) task group is tackling the definition of an alternate UWB radio physical layer for the 802.15.3 W-PAN standard [Gilb 2003]. The IEEE 802.15.4 project in a new group 4a, along with the ZigBee Alliance [Zig-Bee 2003], is considering a new physical layer to be the standard for the purpose of including positioning and distancing capabilities. The Zig-Bee Alliance is an association of companies working together to enable reliable, cost-effective, low-power, wirelessly networked monitoring and control products based on an open global standard. UWB proposals within 4a are inevitable because of the demonstrated positioning capabilities of UWB systems (see [IEEE802 03/157]). A portion of the IEEE 802 organization is shown in Figure 3.1.

From the point of view of standards activities, the IEEE Project 802.15.3a task group provides us with examples of three very different UWB radio systems. All three systems described here evolved from the Project 802.15a UWB PHY (Physical Layer) proposal process. As of this writing, the approach to a standard has not been chosen, and perhaps this reflects the abundance of good choices! The three implementations show how different a physical description of a radio can become while conforming to the same broad regulatory definitions seen in Chapter 2. One method is based on Orthogonal Frequency Division Multiplexing (OFDM) and another is based on Direct Sequence Ultrawideband (DS-UWB). The third method is a Time Division/Frequency Division Multiple Access (TD/FDMA) pulse approach. Of these, the TD/FDMA and DS-UWB are closest to an impulse radio, and their wavelet

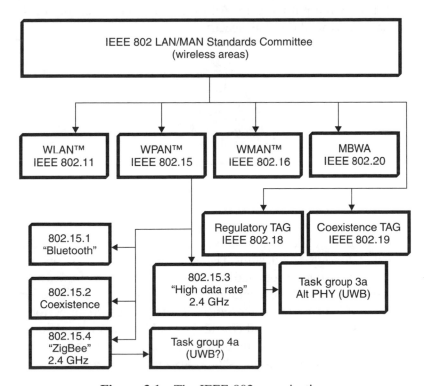

Figure 3.1   The IEEE 802 organization.

characteristics will be studied in detail when we consider UWB signal design in Chapter 4. Other than a brief overview, it is not our intent to describe or analyze either OFDM or DS-CDMA, but rather to show how these well-known radio techniques have been applied to conform with the regulatory definitions for access to the UWB spectrum.

The 802.15.3a process has exposed several viable techniques to access the UWB spectrum. Three of these are briefly summarized in Table 3.1. They are "OFDM" (see [IEEE802 03/449]), "DS-CDMA" (see [IEEE802 03/334]), and "TD/FDMA" pulses (see [IEEE802 03/109]). All three systems access a minimum of about 1.5 GHz in the simplest mode of operation. OFDM and TD/FDMA systems hop among at least three channels of about 500-MHz width within the 3.1- to 5-GHz range. DS-CDMA uses one channel of 1.4-GHz bandwidth. All three have a way of utilizing the segment of spectrum between 3.1 and approximately 5 GHz with remarkably similar performance. All of them, additionally, have ways of increasing performance, again essentially by similar amounts, by utilizing the rest of the UWB up to 10.6 GHz. Table 3.2 shows that each method delivers more than 100 Mbps at 10-m distance and well over 400 Mbps at 4 m with a 6-dB margin to spare. All of the methods meet the regulatory criteria for access to the UWB spectrum.

An Abundance of Good Choices

Modulation efficiency is the needed energy per bit to noise density ratio for achieving a bit error rate (BER) no greater than a specified value. A BER less than 10^{-3} is used for comparison in Table 3.1. The "margin" values in Table 3.1 should be used with a great deal of caution. Table 3.1 is not meant for system design. The margin is calculated for an ideal condition of free-space propagation and is equivalent to a measurement in a well-isolated anechoic chamber. It is a suitable and stable way of *comparing* different systems. Designers wishing to estimate performance in realistic environments should consider the additional propagation losses usually encountered in other than ideal environments. Performance in multipath and shadowing conditions should also be considered.

Table 3.1 Three ways of implementing UWB systems that meet regulatory requirements.

	OFDM	DS-UWB	TD/FDMA pulses
Bands	3–13	2	3–13
Bandwidths	3 × 528–13 × 528 MHz	1.5 and 3.6 GHz	3 × 550–13 × 550 MHz
Frequency ranges, GHz	3.1–4.8 4.8–10.6	3.1–5.15 5.825–10.6	3.1–5 4.9–10.6
Modulation	OFDM-QPSK	M-BOK, QPSK	M-BOK, QPSK
Modulation efficiency: 10^{-3} BER	6.8 dB	4.1–6.8 dB	6.1–6.8 dB
Error correction	Convolutional	Convolutional and Reed–Solomon codes	Convolutional and Reed–Solomon codes
Margin at 10 m	6 dB at 110 Mbps	6 dB at 112 Mbps	6 dB at 108 Mbps
Margin at 4 m	11 dB at 200 Mbps	11 dB at 224 Mbps	8 dB at 288 Mbps
Margin at 4 m	6 dB at 480 Mbps	6 dB at 448 Mbps	4 dB at 577 Mbps

Table 3.2 OFDM system parameters for a UWB system.

Information data rate	110 Mbps	200 Mbps	480 Mbps
Modulation/constellation	OFDM/QPSK	OFDM/QPSK	OFDM/QPSK
FFT size	128	128	128
Coding rate	11/32	5/8	3/4
Spreading rate	2	2	1
Information tones	50	50	100
Data tones	100	100	100
Information length	242.42 ns	242.42 ns	242.42 ns
Cyclic prefix	32/528 = 60.61 ns	60.61 ns	60.61 ns
Guard interval	5/528 = 9.47 ns	9.47 ns	9.47 ns
Symbol interval	312.5 ns	312.5 ns	312.5 ns
Channel bit rate	640 Mbps	640 Mbps	640 Mbps
Symbol period	937.5 ns	937.5 ns	937.5 ns

3.1.1 An OFDM Approach to UWB

OFDM technology is four decades old. It appears in many communications services, both wired and wireless. Asymmetric Digital Subscriber Live (ADSL) services use OFDM over "plain old telephone system" (POTS) wire lines. OFDM also appears in the IEEE 802.11a/g and IEEE 802.16a wireless local area network (WLAN) standards, in Digital Audio Broadcasting (DAB), as well as in the digital terrestrial television broadcast systems: DVD in Europe and ISDB in Japan. The OFDM approach to UWB meets the "500-MHz bandwidth" requirement by using 128 carriers that are modulated with a Quadrature Phase Shift Keying (QPSK) constellation. The composite signal occupies a 528-MHz wide channel. The OFDM carriers are efficiently generated using digital Fast Fourier Transform (FFT) techniques. The OFDM composite signal persists on a channel for the information length of 242.42 ns plus the 60.61-ns cyclic prefix time, and then switches to another channel within a 9.5-ns guard time. The long dwell time on a channel, approximately 303 ns, means that multipath fading is effectively viewed through a narrowband filter. This OFDM system will encounter full Rayleigh fading in multipath channels. Because the UWB emission regulations limit the power per megahertz, a larger utilized bandwidth will result in a larger total emitted average power. This OFDM approach uses 528-MHz channels, but uses at least three at a time in a time-frequency hopping manner, thus emitting the total power allowed in 1,584 MHz. Frequency hopping is used in this OFDM approach as a way of having this system occupy a large total bandwidth for the purpose of increasing the total radiated power. Frequency hopping, especially as a form of spread spectrum for secure communications, dates back to the Hedy K. Markcy (Hedy Lamarr) and George Antheil patent [Markey 1942].

The system parameters for an OFDM system are shown in Table 3.2. The system supports at least three data rates: 110, 200, and 480 Mbps, with additional data rates possible (see, for example, the IEEE 802.15.3a proposal [IEEE802 03/449]). QPSK modulation is used on the OFDM tones and a 128-point Fourier transform generates the OFDM tones. The different data rates are supported by different combinations of error-correction coding rates, the number of tones carrying data, and the spreading rate on each tone. Additional 528-MHz wide channels all the way up to the band limit at 10.6 GHz provide additional system capacity, more available power, and combinations of power and capacity to further improve performance. Once again, it is not our intent to provide design details of this

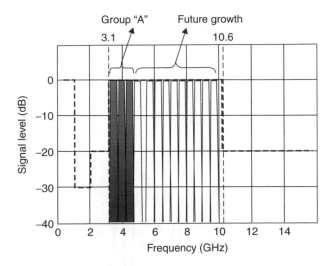

Figure 3.2 Band occupancy of a frequency-hopping OFDM UWB system.

approach, but rather to expose that, with the proper choice of parameters, a rather traditional radio approach can be configured to meet the regulatory requirements for access to the UWB spectrum. Figure 3.2 shows the band plan for the proposed frequency-hopping OFDM system. Minimal systems would use the "A" channels shown in the figure. Additional channels beyond this group are available for more elaborate systems and for future growth.

This OFDM approach uses conventional radio techniques to access the 7,500 MHz of unlicensed spectrum under the UWB rules. Is it "UWB?" Yes! The regulations do not specify technology, but rather rules for accessing spectrum. Is it "traditional UWB?" Perhaps not. The performance of this OFDM will resemble that of 128 narrowband radios, each occupying about 4 MHz of spectrum. Multipath is dealt with by spreading the forward error-correction code across the frequency bands. There is not much depth in the coding at the higher data rates, so comparatively less protection is available at higher rates.

3.1.2 A DS-UWB Approach to UWB

Direct Sequence Spread Spectrum (DS-SS) has roots in secure and military communications systems. It also appears in the IEEE802.11b WLAN standard. We had encountered DS-SS earlier when we discovered that

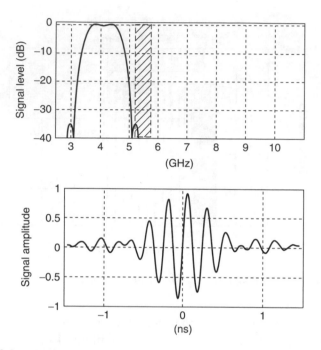

Figure 3.3 UWB wavelet and its spectrum in the 3.1- to 5-GHz range.

the regulatory allocation of a block of spectrum to direct-sequence code division multiple access (DS-CDMA) was another of the important regulatory changes in spectrum management. Users of the spectrum could simultaneously occupy a large common channel and be separated by digital codes rather than by the wavelength of frequency. If the spreading is "noiselike" enough, Shannon predicted that, with the right conditions, the spectrum could be most efficiently utilized [Shannon 1948]. This is precisely the premise behind the DS-UWB approach to a UWB system intended for the UWB spectrum (see [IEEE802 03/334]). In a specific implementation summarized in the DS-UWB column of Table 3.1, we see direct-sequence approach designed to occupy at least 1.5 GHz bandwidth in the 3.1- to 5.15-GHz range and 3.7-GHz bandwidth in the 5.8- to 10.6-GHz range. Unlike a traditional carrier-based DS-SS system, this UWB approach uses nonsinusoidal wavelets tailored to occupy the desired spectrum in an efficient manner. Figure 3.3 shows a sample wavelet along with its spectral content in the 3.1- to 5.1-GHz band. Figure 3.4 shows another sample wavelet, which occupies the 5.8- to 10.6-GHz band. The lowest cost systems are envisioned to use the lower frequency band. The

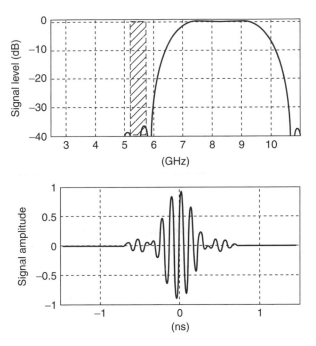

Figure 3.4   UWB wavelet and its spectrum in the 6- to 10.6-GHz range.

two bands can be used independently, or together, to provide a range of options for system deployment. This system uses a Multi-level Bi-Orthogonal Keying (M-BOK) modulation or M-BOK in combination with QPSK depending on the required data rate.

The UWB bandwidth spreading is accomplished by sending the wavelets shown in Figures 3.2 and 3.3 at the 1.368-Gcps chip rate. Modulation of the sequence of wavelets "whitens" the spectrum, which means that it disrupts regularities that would otherwise result in spectral lines. M-BOK modulation comprises 24- and 32-length ternary orthogonal sequences $(-1, 0, +1)$ forming symbol wavelets. Either 1, 2, 3 or 6 bits are sent with each code symbol. For example, a 64-BOK modulation carries 6 bits at a time ($2^6 = 64$ BOK combinations) to form a symbol. The 24-length ternary codes are used with 2-BOK, 4-BOK, and 8-BOK, while 32-length codes are used with 64-BOK. M-BOK modulation has the property that as M increases without bound, the modulation efficiency approaches the Shannon-limited value of -1.59 dB.

Symbols are sent at 42.75 MSym/s resulting in a channel bit rate of 256.5 Mbps. The data are encoded with a rate 0.44 error-correction code

Table 3.3 One possible set of DS-UWB system parameters for the 3.1- to 5.1-GHz band.

Information data rate	112 Mbps	224 Mbps	448 Mbps
Modulation/constellation	64-BOK	QPSK/64-BOK	QPSK/64-BOK
Symbol rate, MSym/s	42.75	42.75	42.75
Coding rate	0.44	0.44	0.87
Code length chips/s	32	32	32
Channel chip rate	1.368 Gcps	1.368 Gcps	1.368 Gcps

resulting in the 112-Mbps information data rate seen in Table 3.3. A second set of wavelets that have orthogonal properties to the wavelets pictured in Figures 3.3 and 3.4 are sent on top of the original wavelets in a manner analogous to QPSK in carrier-based radio systems. With QPSK, there are up to 12 bits per symbol. This doubles the data information rate to 224 Mbps. Finally, a 0.87 rate error-correcting code is used with QPSK to achieve 448-Mbps data information rate. Many other combinations of modulation depths (M-BOK) and coding rates are possible. The M-BOK codes, along with forward error correction, are especially effective in multipath propagation. This example of a UWB system uses wavelets that are approximately 1.5 GHz and 3.6 GHz, respectively, when measured at the points 10 dB down from the peak level. They resemble an impulse approach in which the impulses are sent with minimal spacing between them. The design and characteristics of such wavelet impulses are covered in Chapter 4.

3.1.3 A TD/FDMA Approach to UWB

The TD/FDMA Access approach uses impulses that are centered at frequencies spaced by 550 MHz. In the approach (see [IEEE802 03/109]), the impulses are 3-ns long and occupy a 700-MHz bandwidth when measured at points 10 dB below the peak value. The impulse wavelets for the lower four bands are shown in Figure 3.5. The corresponding spectral occupancy is portrayed in Figure 3.6.

In this approach, the multipath is mitigated by the time interval between pulses that appear on the same channel. The TD/FDMA system uses impulses that are separated in time. The design and characteristics of such pulses are covered in Chapter 4.

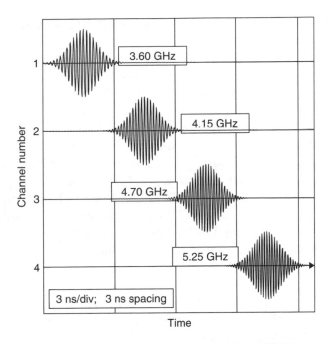

Figure 3.5 Impulse wavelets for a TD/FDMA UWB system.

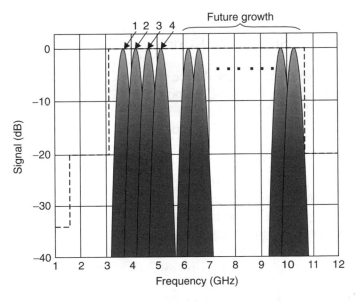

Figure 3.6 Spectrum occupancy of the TD/FDMA UWB system.

3.2 Positioning and Location in UWB Standards

UWB technology is uniquely positioned to provide distancing and location capabilities to wireless sensor networks. The popularity of wireless sensor networks is expected to grow dramatically in the next few years, because they provide practical machine-to-machine communication at a very low cost. The growth is expected to duplicate the recent explosion of wireless LANs (see [Callaway 2003]). Accurate distance measuring using UWB has been demonstrated adequately in recent years, and several corporations had developed chip sets as early as 1993, demonstrating distance-measuring capabilities (see [AetherWire 2003], and [Time 2003]). Aether Wire & Location, Inc., chose a DS-SS approach (see [Fleming 1998]) to measure distance using UWB. Time Domain Corporation pioneered a Time-Modulated Impulse approach (see [Fullerton 1989]), which not only measured position but also fused data communications within the same protocol.

In the IEEE 802 organization, a new group, 4a, has begun work to propose a new radio physical layer for the 802.15.4 Standard [IEEE802.15.4]. Because of the unique capabilities of UWB to satisfy the need for positioning, low data rates at longer distances, and high system capacities, UWB proposals are inevitable.

3.3 European Standards Efforts

"Standards" take on a slightly different connotation in Europe. There are two European groups, CEPT SE24 and ETSI TG31a, with primary responsibility for UWB matters. CEPT SE24 (European Conference of Postal and Telecommunications Administrations Spectrum Engineering Committee 24) is currently studying UWB frequency sharing with other services in the band 3.1 to 10.6 GHz. Results of this committee will provide guidance to European regulators and provides European input to TTU-R TG 1/8. ETSI TG31a are focusing on conformance testing standards in support of the SE24 study effort. In the European process, ETSI TG31 used the US FCC UWB spectrum emissions mask specification as a starting point and modified it by adding sloping out-of-band skirts (see Figure 2.6). Completion of work may be waiting for 802.15.3a to decide on modulation and waveforms.

In February, 2003, the Directorate General of the European Commission adopted a mandate to "Harmonize standards covering Ultrawide band (UWB) applications," [ETSI 2003]. The purpose of the mandate is to establish a set of standards covering UWB applications to be recognized under Directive 1999/5/EC (the "R&TTE Directive"), giving a presumption of conformity with its requirements. This mandate defines the essential requirements the R&TTE equipment must meet to be placed on the market and to be put into service for its intended purpose.

In Europe, UWB is generally understood to be a technology, which, by transmitting exactly timed pulses, spreads transmitted electromagnetic energy over a very large frequency range, with the result that the spectral power density lies below the classical electromagnetic compatibility (EMC) limits. Note that this is an "impulse radio" sort of definition, rather than the minimum-bandwidth definition relied upon in the United States. By comparison, these limits amount to the -41.3 dBm/MHz in the UWB band adopted in the United States. Proposed applications of the UWB technology include communications, anticollision radar, and imaging techniques. Proponents argue that UWB devices can operate without causing interference to other users of the spectrum. As a technology, UWB thus does not fit in the classical radio regulatory paradigm, which bases itself on a subdivision of the spectrum in bands that are then allocated for specific uses. Incumbent spectrum users have voiced concerns that the cumulative effect of UWB devices raises the background noise for their spectrum, rendering the operation of their services difficult or sometimes even impossible. Public authorities, especially in the United States and in Europe, are studying these effects. The European Communications Committee held several workshops on the matter. Such studies should lead to the formulation of specific protection requirements for critical services to be taken into account in harmonized standards for UWB devices.

The European Standardization Organizations are mandated to

- develop a work program for harmonized standards covering UWB applications;

- report the progress of the work to the Commission at regular intervals and at least prior to each meeting of the Telecommunications Conformity Assessment and Market (TCAM) Surveillance Committee, which is the standing committee set up by the Directive;

- deliver harmonized standards for the work items confirmed by the TCAM, the references of which will be published in the official journal of the European Communities as giving presumption of conformity with the R&TTE Directive.

The Directorate General recommends that experts should liaise intensively with regulatory bodies and their experts. The delivery of harmonized standards dealing with UWB is expected as of December 2004. Where appropriate, equivalent activities in the ITU and in ISO/IEC should be aligned. Due account should be taken of regulations and draft regulations adopted in other economies so as to ensure a global market for UWB devices.

3.4 Summary

Standards make possible the wide acceptance and dissemination of products from multiple manufacturers and vendors with an economy of scale that reduces costs to consumers. Currently, only the IEEE organization and ETSI are developing standards related to UWB. The IEEE, under the 802 family of standards, is developing a high data rate radio physical layer to the 802.15.3 W-PAN standard. In Europe, ETSI are developing standards for conformance tests in support of spectrum-sharing studies that will result in European UWB regulations. The work is gauged somewhat by the outcome of the IEEE 802 process, which will define waveforms and modulations. The response to the IEEE 802.15.3a call for proposals generated a wide range of UWB implementations, including impulse radios, direct-sequence approaches, and narrowband radios like OFDM having parameters that extend the frequency bandwidth to beyond the 500-MHz minimum required under the regulations. UWB has potential for capabilities that are not easily done in narrow bandwidths such as accurate distance measurements. These will drive the additional future standardization efforts.

References

[AetherWire 2003] Aether Wire & Location, Inc., (Online): <http://www.aetherwire.com/> 7 December 2003.

[Callaway 2003] E. H. Callaway, *Wireless Sensor Networks: Architectures and Protocols*, New York: Auerbach Publications (CRC Press), 2003.

[ETSI 2003] ETSI, *Harmonised Standards Covering Ultrawide Band (UWB) Applications*, Directorate General of the European Commission, Standardisation Mandate: DG ENTR/G/3M/329, Brussels, 25 February 2003.

[Fleming 1998] R. A. Fleming and C. E. Kushner, *Spread Spectrum Localizers*, U.S. Patent 5,748,891, 5 May 1998.

[Fullerton 1989] L. Fullerton, *Time Domain Transmission System*, U.S. Patent 4,813,057, 14 March 1989.

[Gilb 2003] J. K. Gilb, *Wireless Multimedia: A Guide to the IEEE 802.15.3 Standard*, NJ: IEEE Press, 2003.

[IEEE802 03/109] IEEE802 Document: [03109r1P802-15_TG3a-Intel-CFP-Presentation.ppt].

[IEEE802 03/157] IEEE802 Document: [03157r1p802-15_wg-understanding_uwb_for_low-power_communications-a_tutorial.pdf].

[IEEE802 03/334] IEEE802 Document: [15-03-0334-05-003a-xtremespectrum-cfp-presentation.pdf].

[IEEE802 03/449] IEEE802 Document: [15-03-0449-03-003a-multi-band-ofdm-physical-layer-proposal-update.ppt].

[IEEE802.15.4] IEEE Std 802.15.4™-2003 Part 15.4: *Wireless Medium Access Control (MAC) and Physical Layer (PHY) Specifications for Low-Rate Wireless Personal Area Networks LR-WPANs)*, 1 October 2003.

[Markey 1942] H. K. Markey and G. Antheil, *Secret Communication System*, U.S. Patent 2,292,387, 14 April 1942.

[McKeown 2003] D. McKeown, *Gammz UWB Cartoons and Art*, Private Communication to K. Siwiak, December 2003.

[Shannon 1948] C. E. Shannon, "A mathematical theory of communication," *The Bell System Technical Journal*, **27**, 379–423, 623–656, 1948.

[Time 2003] Time Domain Corporation, (Online): <http://www.time domain.com>, 17 December 2003.

[ZigBee 2003] ZigBee Alliance, (Online): <http://www.zigbee.org/>, 17 December 2003.

4

Generating and Transmitting UWB Signals

Introduction

What can a UWB signal look like in a world without spectrum constraints? UWB might have taken on the character of "pure impulse radio" with signal energy spread over vast amounts of spectrum. UWB signals would be simple impulses. Unfortunately, there are constraints. The frequency spectrum is already fully allocated. There is no "new" spot where UWB can flourish. Instead, early proponents of the technology recognized that UWB could be overlaid on already occupied spectrum. The idea could be successful only if the UWB signal levels were set to emission levels below what existing spectrum users considered tolerable "unintentional" emissions.

Many electrical and electronic devices radiate noise as a by-product of normal operation. Regulatory agencies seek to restrict these emission levels to very low levels, which are presumed to be nonharmful to existing licensed radio services. It is these levels that UWB proponents targeted as levels for the intentional radiation of UWB signals. Initial experiments were conducted in spectrum above about 4 GHz. Signals were generated by simple "pulsers" supplying wideband energy to wideband antennas.

The first step in a radio communication link involves the generation of a suitable signal, which is then modulated with desired information. Several factors decide what a UWB signal needs to look like. These factors include

1. what is permissible under the rules and regulations;
2. what level of coexistence with other services in the band is desirable;

Ultra-Wideband Radio Technology   Kazimierz Siwiak and Debra McKeown
© 2004 John Wiley & Sons, Ltd   ISBN: 0-470-85931-8

3. what technological constraints there are from the feasibility, cost, and
marketability points of view.

Rules and regulations such as those of the FCC do not define UWB
technology. The regulations define the broad rules and conditions for
UWB communication systems to access and share a 7,500 MHz swath
of spectrum extending from 3.1 to 10.6 GHz. To gain market acceptance, UWB radios will need to coexist with, share, and even perhaps
interoperate with other radio services. Finally, UWB technology will
need to be physically implemented in cost-effective integrated circuits,
which themselves have various limitations in performance across the frequency band. Against these constraints we now seek to generate useful
and cost-effective UWB signals that are permissible under existing regulations.

By way of example, one strategy in generating useful UWB signals
can begin by selecting certain broad design parameters, here based on
US FCC regulations. One parameter, the lower operating frequency of
3.1 GHz, is dictated by such regulations. A practical upper limit might
be based on technical capabilities of integrated circuit implementations as
implied in Figure 4.1 (near 6–7 GHz for some of the lower cost-integrated
circuit processes) and on coexistence parameters with other services such
as the U-NII band services operating between 5.15 and 5.825 GHz in the

Figure 4.1   Some low-cost integrated circuit processes "run out of steam" near
6 GHz [McKeown 2003].

United States, and the 4.9 to 5.091 GHz license-free bands under development in Japan. These services include IEEE 802.11a compliant data devices that could be part of a data infrastructure in a local area network, which also works in conjunction with the UWB radios colocated within the same devices. The desired operating band is therefore between 3.1 and approximately 5 GHz. Figure 4.2 illustrates, in broad terms, some of the design constraints on UWB systems. The FCC regulations at 3.1 GHz form a "hard limit" in that the UWB signal must be 20 dB below the peak permissible emission level of −41.3 dBm/MHz defined in the rules. At the upper limit, a "best effort" approach is used to keep the UWB emissions from interfering with, or being interfered, by the services operating above 4.9 GHz. Thus, the broad description of the UWB signal subjected to the constraints is a signal wider in bandwidth than 500 MHz (FCC regulation) and no wider than about 2 GHz extending from 3.1 to about 5 GHz. A wide range of suitable signal designs is possible within the bounds of our example. Several such signal designs can be described.

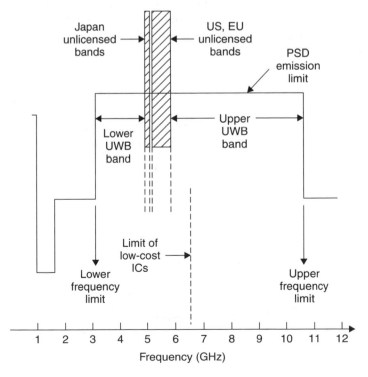

Figure 4.2   Some constraints for UWB systems.

4.1 UWB Signal Definitions

Regulations (in this example, FCC regulations: see Appendix A) shape the general characteristics of UWB signals and signaling systems. Our primary concern is with *UWB Communications and Measurement Systems*, such as high-speed home and business networking devices. The devices must operate in the UWB frequency band 3.1 to 10.6 GHz. The equipment must also be designed to ensure that operation can only occur indoors or it must consist of handheld devices that may be employed for such activities as peer-to-peer operation. Intentional UWB radiators must be designed to guarantee that the 20-dB bandwidth of the emission is contained within the UWB frequency band. The minimum bandwidth measured at points 10 dB below the peak emission level is 500 MHz. The permissible emission levels for UWB signals in the UWB band are set at −41.3 dBm/MHz.

Therefore, in UWB signal design, we are concerned with the 10-dB bandwidth to ensure compliance with minimum bandwidth requirements, the 20-dB bandwidth to make certain that the signal remains below the 20-dB band edge corners on the UWB communications power spectral density (PSD) mask, and the frequency of highest radiated emission, which must be below the maximum allowed PSD. Figure 4.3 represents these limits pictorially.

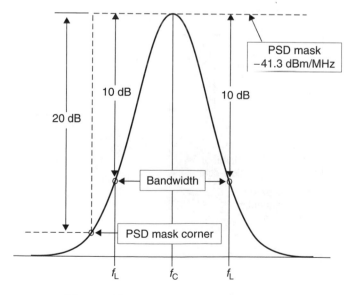

Figure 4.3 UWB signal design points.

4.2 Approaches to Generating UWB Signals

UWB signals may be generated by a great variety of methods. One might choose to supply narrow impulses to a band-pass filter, as had been done in the earlier UWB experiments. Alternatively, one could generate precise UWB signal shapes and accurately place them in the allowable spectrum. Finally, one can simply scale the modulation methods of conventional radio systems, such as direct sequence spread spectrum (DSSS) systems [Aetherwire 2003, IEEE802 03/334] or orthogonal frequency domain multiplexed (OFDM) systems [IEEE802 03/449], so that the bandwidth resulting from the modulation occupies at least the minimum regulatory amount. Figure 4.4 shows a conceptual rendition of a *"retro-UWB"* system based on impulse excited filters, which can be configured to meet the regulatory definition of UWB. The example (see [IEEE802 03/157]) is not meant to be an example of a practical UWB system, but rather shows the close tie between modern concepts in UWB and the roots of early wireless, as traced in Chapter 1.

In this modern flavor of an early wireless spark system, the "transmitter" of Figure 4.4 comprises a digital data source, which supplies impulse energy at the input data rate to a "resonant" Tesla-like transformer and spark-gap apparatus. The "Q" or "sharpness" of the resonant circuit is selected so that the resulting emission bandwidth is at least 500 MHz. The receiver is a simple amplitude level detector followed by a data filter. While more tongue-in-cheek than practical or efficient, this circuit does illustrate, *in principle*, the simplicity of a UWB communications

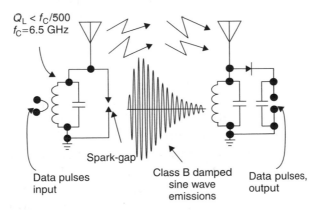

Figure 4.4 Although not especially efficient, and some details need work, this conceptual UWB system reveals early wireless roots.

system and would be permissible under the FCC regulations, which now specifically allow "class B damped sine wave emissions" under the UWB rules. Modern approaches [Fullerton 1989] to UWB systems based on impulse excited filters employ efficient coherent receivers and precise pulse positioning resulting in efficient data communications systems and accurate positioning and radar devices. In this chapter, we will concentrate on the principles behind precision UWB pulse generation.

4.2.1 UWB Signal Design

Some design choices for UWB signals are spawned by regulations. The permissible emission levels (in the US) for UWB signals are set at -41.3 dBm/MHz, so the total radiated power depends directly on the amount of spectrum utilized. Higher levels, -35.3 dBm/MHz, are permitted under experimental licensing in Singapore's UWB Friendly zone, and levels as low as -61.3 dB/MHz are being considered in Europe. This means that

1. we need to use as much bandwidth as possible to maximize the emitted power;

2. because of the antenna aperture scaling, the lower frequencies in the UWB band are the most valuable for our purposes.

Signals are designed by shaping their properties as a function of time. There is a one-to-one mapping between what a signal looks like in time and its frequency spectrum. This relationship is mathematically expressed by the Fourier transform. The key point here is that once the signal is fully specified or described in one domain (time or frequency), its properties in the other domain are given by the Fourier transform. We use this property to define the signal in one domain and to extract the desired properties in the other domain. One way of generating UWB signals involves using very sharp signal transitions in time, like a step function, or a very narrow rectangular pulse, followed by a band-shaping filter. "Step functions" are the leading and trailing edges of base band digital pulses.

Sharp signal transitions and extremely narrow pulses (see Figure 4.5) act like extremely wideband energy sources, which are then shaped by the desired band-pass filter. The filter, in effect, "rings like a struck bell" in response to the impulse input. The impulse evokes the transient response of the filter. The output signal is no longer a step or a narrow pulse but rather a complex elongated signal in time, which represents the impulse response of the filter – the "ringing tone" of a bell.

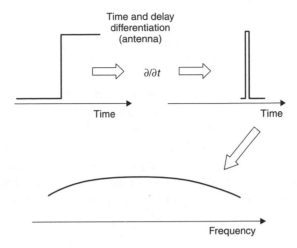

Figure 4.5   Sharp signal edges can generate UWB signals.

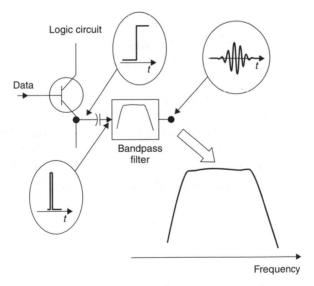

Figure 4.6   Output from a logic circuit generates UWB signals, which can be filtered to appear in the desired band.

Supplying a sharp impulse to the input of a band-pass filter is a simple and effective method of generating UWB signals (see Figure 4.6). While appropriate for certain kinds of UWB systems, the method lacks the finesse to precisely shape and position UWB signal energy in the frequency spectrum.

4.2.2 Precision Signal Design

In another approach, we can synthesize UWB signals with precision in both the signal shape and its placement in the frequency spectrum. To that end, signals can be shaped at "base band" and then shifted in frequency to the desired location in the spectrum using Armstrong's heterodyning technique. Band shaping is much easier to accomplish at base band than it is at higher radio frequencies. We first choose a signal shape so that the bandwidth specifications are met. We will consider several examples of base band pulse shapes starting with a simple rectangular shaped signal representing, perhaps, a digital data stream in the time domain. The rectangular pulse $r(t)$ is centered on $t = 0$ and has a width of T picoseconds.

$$r(t) = 1, \left\{ -\frac{T}{2} < t < \frac{T}{2} \right\}; \quad 0 \text{ elsewhere} \tag{4.1}$$

The Fourier transform of $r(t)$ gives us an expression $R(f)$ for the frequency-domain representation of the same signal,

$$R(f) = \frac{T \sin(\pi T f)}{\pi T f} \tag{4.2}$$

Figure 4.7 shows the relationship between $r(t)$ and $R(f)$ graphically. Recall from our discussions in Chapter 1 that the narrower a signal is in time, the wider its representation in the frequency spectrum. That behavior is captured in the scaling parameter T in $r(t)$ and $R(f)$. Notice that the width T in time corresponds to a width $2/T$ in frequency. We now have a way of selecting the width of the signal spectrum in terms of the pulse duration.

We see in Figure 4.7 that the rectangular pulse generates a frequency-domain signal having a large main lobe between $f = -1/T$ and $f = 1/T$, but that it also has significant energy lobes outside the main lobes. These are troublesome since they carry energy at levels that might exceed regulatory levels. We need an effective way to lower these side-lobe levels.

An interesting property of signals is that the smoother the rise and fall of a time-domain signal, the lower the energy content in the side lobes of the frequency representation of the signal. Thus, by shaping the edges of a time signal, we can control the energy content in the spectral side lobes. Let us look in detail at three kinds of pulse shapes:

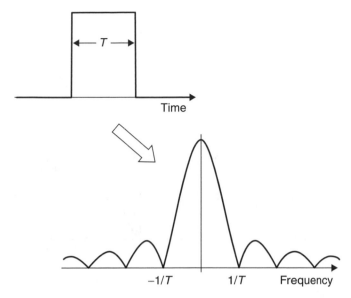

Figure 4.7 Time-domain and frequency-domain representations of a pulse are related by the Fourier transform.

(a) a rectangular pulse $r(t)$

(b) a cosine-shaped pulse $c(t)$, and

(c) a bell shaped-Gaussian pulse $g(t)$.

These signals are depicted by Figure 4.8 in both the time and frequency domains. The cosine pulse, described by

$$c(t) = \cos\left(\frac{2\pi t f_a}{2}\right); |t| < \frac{1}{2f_a}; \quad 0 \text{ elsewhere} \qquad (4.3)$$

rounds the top part of the rectangular pulse, but still has abrupt corners at the two points where $c = 0$. Its frequency-domain representation is

$$C(f) = \frac{\cos\left(\pi \frac{f}{f_a}\right)}{1 - \left(2\frac{f}{f_a}\right)^2} \qquad (4.4)$$

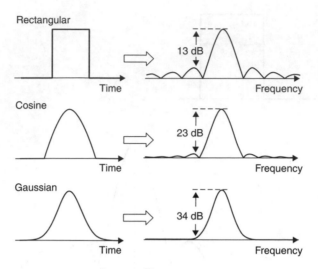

Figure 4.8 Smoother time-domain signals have less energy outside the desired frequency band.

The bell-shaped Gaussian pulse $g(t)$, written for graphing convenience with its peak value set to 1, is described by

$$g(t) = \exp\left(\frac{-0.5\,t^2}{u^2}\right) \tag{4.5}$$

and its frequency representation, also a Gaussian shape, is

$$G(f) = \exp[-2(\pi f u)^2] \tag{4.6}$$

The Gaussian time pulse has smooth transitions everywhere, and similarly is represented by a smooth bell shape in frequency. The width parameter $u = u_B$ is picked so that $G(f)^2 = 0.1$ at a value f_B GHz to meet the bandwidth requirements as seen in Figure 4.3. So,

$$u_B = \frac{1}{[2\pi f_B[\log(e)]^{1/2}]} \tag{4.7}$$

and f_B is the desired half bandwidth of $1\,\mathrm{GHz}$ and $e = 2.81828\ldots$ is the base of the natural logarithms. Design parameters for the three time pulse shapes, rectangular, cosine, and Gaussian, can be chosen so that the 10-dB bandwidths in the frequency-domain representation are exactly

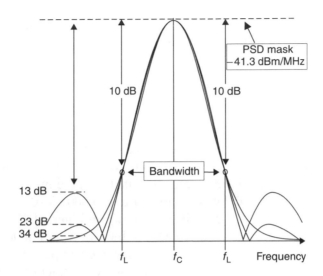

Figure 4.9 Different time-domain signals have the same bandwidth, but less energy outside that bandwidth.

equal, as shown in Figure 4.9. The main lobes between the 10-dB points are nearly identical, but the side-lobe structure is quite different. Very clearly, the sharpest edged time pulse has the highest side-lobe level, followed by the smoother cosine pulse. The lowest side-lobe, or out-of-band, energy is contained in the spectrum of the Gaussian pulse. While only the fundamental basics of pulse design are illustrated here, we should point out that many more and different pulse designs are possible, and in fact, desirable in UWB systems. For example, pulse shapes whose spectra conform closer to the regulatory emissions mask would advantageously contain more energy.

4.2.3 Calculating Power for Repetitively Sent Pulses

A single isolated pulse delivers *energy*, measured in joules. Energy pulses are delivered at an average signal pulse repetition frequency (PRF), so they appear as energy per unit time (joules/s), or power. The emission rules and regulations specify limits on both average power emissions and peak power. Average power emission is measured as effective isotropically radiated power (EIRP) – that is, power radiated by an antenna with a gain of 1. Absolute peak power is measured over the duration of the pulses that are sent repetitively. However, for purposes of conforming to regulations, peak power is the power measured in a 1-MHz bandwidth. The total emitted power for a repetitively sent signal $S(f)$ whose peak PSD exactly equals the limit -41.3 dBm/MHz is found by

$$P_{\text{EIRP}} = 10 \ \log \left(\int_0^\infty S(f)^2 \, df \quad 10^{-41.3/10} 10^{-6} \right) \qquad (4.8)$$

The peak amplitude value of $S(f)$ had been set to 1, and the frequency f is in hertz. The closer that $S(f)$ conforms to the area under the PSD limits, the more power the signal can be made to transmit. If all available spectra from 3.1 to 10.6 GHz were perfectly filled with the maximum allowed signal PSD, the total EIRP would amount to $(10,600-3,100) \ 10^{-41.3/10} = 0.556$ mW, equivalent to -2.55 dBm as shown in Figure 4.10. This represents the absolute maximum possible EIRP limit for UWB under these particular regulations. The figure also shows the average power, -10.6 dBm, represented by 2-GHz wide pulses sent at some continuous PRF.

Figure 4.11 shows pulses sent at a particular PRF and illustrates the relationship between pulse shape and spectral bandwidth, but it also shows that pulses sent *repetitively* can generate *regularly spaced lines* in the spectrum. Modulating the pulses, that is, breaking up the pulse regularity

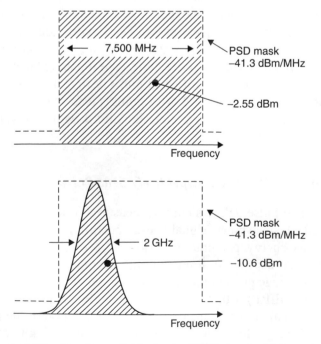

Figure 4.10   Power available in the UWB band and in a signal.

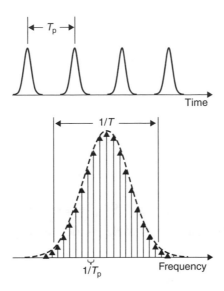

Figure 4.11 Unmodulated pulses sent at a *regular* PRF produce spectral lines.

by coding data as pulse position, or modulating pulses by encoding data as pulse polarity can break up the spectral lines and make the spectrum smoother and more noiselike.

Table 4.1 shows the EIRP for the rectangular, the cosine, and the Gaussian shaped pulses, and also lists the levels of the highest side lobes. The time-domain pulse shapes and their frequency spectra are pictured in Figure 4.8. Note that the design parameters were chosen so that the pulses have exactly the same bandwidth at the level 10 dB down (10% of the peak power spectral density) from the peak level, as seen in Figure 4.9. The signals, thus, have the same bandwidth according to the FCC UWB regulatory definition.

The rectangular pulse produces 0.5 dB more total power than the cosine pulse, and 0.6 dB more than the Gaussian pulse, but it also has the most

Table 4.1 Comparison of total EIRP and side-lobe levels for pulses with 10-dB BW of 2 GHz.

Pulse shape	EIRP (dBm)	Highest side lobe (dB)
Rectangular	−10.0	−13.3
Cosine	−10.5	−23.0
Gaussian	−10.6	−34.5

energy outside the main lobe. The Gaussian pulse does not have side lobes; the energy just continues to decay according to the bell-shaped curve. The value reported in Table 4.1 is the signal level at the frequency corresponding to the peak of the rectangular pulse side lobe. Shaping pulses thus affect the total occupied bandwidth and also the total possible EIRP.

4.3 Signal Pulse Design Examples

Here we will synthesize two different UWB signal designs resulting in

1. a UWB signal that occupies the desired 3.1- to 5.1-GHz segment;

2. multiple signals, each 500-MHz wide, and placed between 3.1 and 10.6 GHz.

In the first design, a pulse is chosen that has a 2-GHz bandwidth, as measured between points that are on either side of the center frequency and are 20 dB below the peak level. The spectrum is to be centered at 4.1 GHz. This pulse is designed to be the widest bandwidth signal that fits between the lower UWB band edge at 3.1 GHz and other wireless communications systems that appear between about 4.9 and 5.9 GHz such as the U-NII services in the United States. In the second design, the emission bandwidth is chosen to be 500 MHz measured at points 10 dB down from the peak emission level. This second signal design is the minimum bandwidth permissible under current FCC regulations. Pulses of this narrower bandwidth design will be placed in various positions between 3.1 and 10.6 GHz. Table 4.2 shows a summary of the two different design goals.

4.3.1 Pulse Design Constraints

Note that the narrowband pulses are specified in terms of the regulatory definitions and constraints. The 500-MHz pulse may be placed anywhere in the allowed spectrum between 3.1 and 10.6 GHz. This narrowband pulse

Table 4.2 Signal pulse design goals.

Pulse type	Bandwidth	Target band
Wideband	2 GHz at 20-dB points	3.1–5.1 GHz
Narrowband	500 MHz at 10-dB points	3.1–10.6 GHz

was designed to have the minimum bandwidth permitted under the regulations, that is, 500 MHz measured at the 10-dB points. The pulse 20-dB bandwidth (see Figure 4.3) must be positioned so that it is within the permissible 3.1 to 10.6 GHz range as seen in Figure 4.2. The 20-dB bandwidth for this pulse is 708 MHz, so the peak frequency must be positioned at least 354 MHz above 3.1 GHz and 354 MHz below 10.6 GHz. Thus f_C must be between about 3.46 and 10.24 GHz. A 500-MHz bandwidth at a 10.24-GHz center frequency corresponds to a fractional bandwidth of less than 5%. This challenges the traditional notion of "UWB," but it is permissible under the US FCC rules and regulations. *The UWB regulations are not about any particular kind of technology or traditions. The regulations spell out the rules under which 7,500 MHz of unlicensed spectrum may be accessed!* This is an important distinction, since the regulations now define how technology will appear in this spectrum and hence in the commercial market place.

The UWB Pulse

4.3.2 Choosing a Pulse Shape

For our example signal design, we choose a Gaussian pulse that has a familiar bell shape defined by Equations (4.5) to (4.7). The Gaussian curve-width parameter u_B is selected so that the "bell" fits appropriately within the design constraints for occupied bandwidth as shown in Table 4.1. The smoothness of the curve will guarantee that the side lobes are low enough to meet regulatory mandates on emissions outside the UWB band.

For the "wideband" 3.1 to 5.1 GHz example, the bell will be centered at 4.1 GHz, and the point intersecting 3.1 GHz, which is 1 GHz to the left of the peak, must be 20 dB down from the peak to meet the regulatory constraint (a factor of 0.1 in voltage amplitude). Similarly, the right side of the bell curve should intersect 5 GHz as low as possible to meet our "best effort" constraint in limiting interference and coexisting with 802.11a services. Thus the bell shaped curve has a maximum at 4.1 GHz, and is 10% of maximum at points 1 GHz to either side of that maximum.

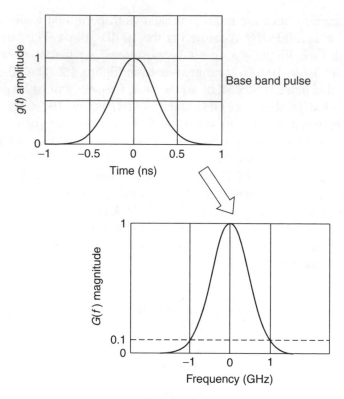

Figure 4.12   Base band pulse and its spectrum.

We state this mathematically using a bell-shaped Gaussian function by Equations (4.5) to (4.7).

Mathematically, the time function $g(t)$ and its frequency representation $G(f)$ are related by the Fourier transform, but we need not dwell on that detail here. We choose a particular u_B so that $G(f) = 0.1$ at $f = 1.0\,\text{GHz}$ to meet the bandwidth requirements as seen in Figure 4.12. The bandwidth f_B in Equation (4.7) is the desired width of $1\,\text{GHz}$. The final step in this UWB signal generation exercise example is to frequency shift (heterodyne) the base band pulse $g(t)$ up to the desired operating center frequency of $4.1\,\text{GHz}$. This comprises a simple mixing with or multiplication by a cosine wave centered at the desired frequency of $f_C = 4.1\,\text{GHz}$. The frequency-shifted signal is

$$g_0(t) = \exp\left(\frac{-0.5\,t^2}{u_B^2}\right)\cos(2\pi f_C) \qquad (4.9)$$

This appears in the frequency spectrum as

$$G_0(f) = \exp[-2(\pi[f \pm f_C]u_B)^2] \qquad (4.10)$$

as seen in Figure 4.13.

The characteristics of $G_0(f)$ meet our design criteria. At 3.1 GHz and also at 5.1 GHz, the signal is 20 dB below (a voltage amplitude factor of 0.1) the peak level.

The maximum allowable EIRP of this pulse is found by integrating the square of $G(f)$ over frequency, and multiplying the result by the peak PSD limit permitted under regulations. In practical designs, some margin must be allowed for spectral peaking due to any regularities in the modulation of the pulses. In the wideband example, $P_{EIRP} = 86\,\mu W$, or $-10.6\,dBm$. Table 4.3 presents a comparison of wide and narrow signal designs. The peaks of the wideband pulses may be located anywhere between 4.15 and 9.55 GHz. This meets the 20-dB emission mask corners at 3.1 and 10.6 GHz. Similarly, the 500-MHz wide pulses may be centered on frequencies between 3.46 and 10.24 GHz.

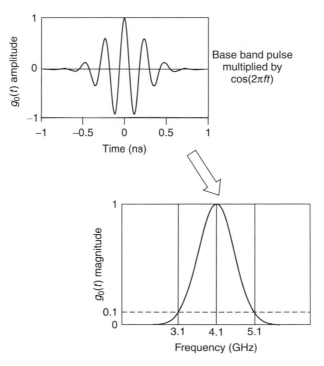

Figure 4.13   Frequency-shifted pulse and its spectrum.

Table 4.3 UWB signal design characteristics for a Gaussian envelope pulse.

Pulse type	Bandwidth	Center frequency	Maximum EIRP
Wideband	2 GHz at −20 dB points	4.15–9.55 GHz	−10.6 dBm
Narrowband	500 MHz at −10 dB points	3.46–10.24 GHz	−16.6 dBm

In fact, multiple such signals, both of the wideband and narrowband design, can be accommodated in the 3.1- to 10.6-GHz band.

4.4 UWB System Band Plans

A possible band plan employing the wideband signal design is shown in Figure 4.14. Five channels, A to E, can be fitted in the 7,500 MHz of spectrum. For systems that need to coexist with various other wireless local area networks (WLANs) operating between 4.9 and 5.9 GHz, channel B can be omitted. Table 4.4 lists the channels and their center frequencies.

A possible band plan for the narrowband signal design is shown in Figures 4.15 to 4.16. Five channels, A to E, can be fitted in the 7,500 MHz

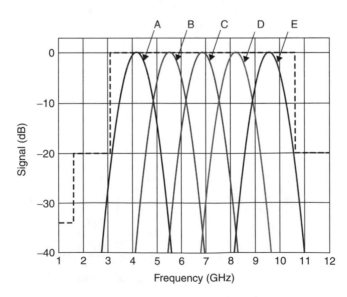

Figure 4.14 A band plan for wideband UWB channels.

Table 4.4   A wideband signal UWB channel plan.

Channel	Center frequency (GHz)	Notes
A	4.15	
B	5.50	WLAN band
C	6.85	
D	8.20	
E	9.55	

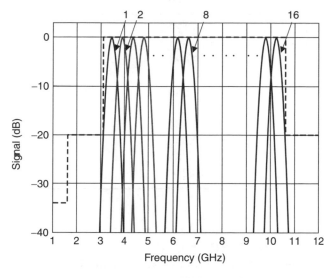

Figure 4.15   A band plan for 500-MHz UWB channels.

of spectrum. Sixteen channels, 1 to 16, can be optionally omitted when there is a need to coexist with various other WLANs operating between 4.9 and 5.9 GHz. Tables 4.4 and 4.5 lists the channels and their center frequencies. The 2-GHz wide signals in Table 4.4 and the 500-MHz wide signals of Table 4.5 represent two of the many possibilities for filling the UWB spectrum with signal energy. Signal impulses may be positioned in spectrum, as shown, and also in time as seen in Figure 4.16.

Tables 4.4 and 4.5 show how pulses of the two candidate designs can appear in the frequency spectrum. They do not address *when* these pulses occupy their respective channels. This is a subject of further design choices that are impacted by factors such as EIRP, data rate, channel multipath characteristics, and IC process maturity. Pulses on various channels may

Table 4.5   A narrowband signal UWB channel plan.

Channel	Center frequency (GHz)	Notes
1	3.460	Global
2	3.912	
3	4.364	
4	4.816	Japan WLAN
5	5.268	
6	5.720	US/EU WLAN
7	6.172	
8	6.624	
9	7.076	
10	7.528	
11	7.980	Global
12	8.432	
13	8.884	
14	9.336	
15	9.788	
16	10.240	

be sent in any variety of time-division multiplexing, so channels may be aggregated to improve EIRP or used to increase system data capacity. There are a wealth of possibilities (see Figure 4.16 for one example) in the choice of pulse bandwidths, center frequencies, pulse repetition frequencies and multiplexing in time. In Figure 4.16, for example, four frequencies from Table 4.5 are used: channel numbers 1 to 3 and either 4 or 5 depending on what the local regulatory and coexistence requirements might be. The pulse on each channel is sent once every 18 ns, allowing moderate multipath to decay. With an average power (see Table 4.3) per channel of -16.6 dBm, a total EIRP of -10.6 dBm may be emitted when an aggregate of four channels is used.

While we show only impulse methods here, many methods and schemes have been proposed in the IEEE 802 Standards forum as candidates for the radio physical layer (PHY) in the IEEE Project 802.15.3a, including "conventional radio" methods like OFDM described in Chapter 3.

We had earlier identified integrated circuit design processes as a consideration in system design choices. Current CMOS processes and techniques might address the design of RF analog circuits with some degree of

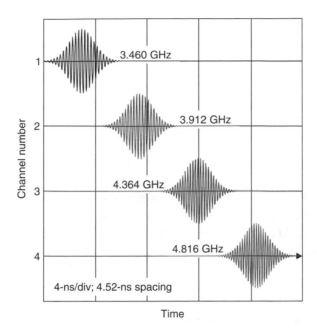

Figure 4.16 A time-multiplexing method for pulses.

economy and facility at frequency ranges up to perhaps 6 or 7 GHz. Thus, cost-effective systems can be designed and deployed to operate, in our example, A to B and 1 to 7 channels. The feasibility of using channels C to E or 8 to 16 might be questionable until IC processes mature more fully. We would expect to see initial UWB products, based on current economical IC processes, to use the lower frequencies initially.

4.5 Overlaying Precision Pulses

Pulses that are orthogonal to each other can be overlaid in both time and frequency without mutual interference. The pulse $g(t)$ was multiplied by a cosine function to shift it to the desired center operating frequency. We will call this the *in-phase signal*, just like the analogous in-phase component in traditional quadrature phase shift keying (QPSK). We could just as easily have used a sine function to accomplish the frequency shift. The resulting pulse in time for this *quadrature* pulse is

$$q_0(t) = \exp\left(\frac{-0.5\,t^2}{u_B^2}\right)\sin(2\pi f_C) \qquad (4.11)$$

which is again analogous to the quadrature term in conventional QPSK signaling. Its frequency equivalent representation is

$$Q_0(f) = j \exp[-2(\pi[f \pm f_C]u_B)^2] \qquad (4.12)$$

Figure 4.17 shows the set of quadrature signals in time as well as the composite sum of the nominal and quadrature signals. The in-phase and quadrature impulses are additionally shown displaced in time by half the duration of the impulse in a manner analogous to conventional Offset-QPSK (O-QPSK) signaling. This offsetting causes the combined signal to have a significantly lower peak-to-average value than if the in-phase and quadrature signals were aligned in time prior to combining.

Both $G_0(f)$ and $Q_0(f)$ occupy exactly the same spectrum, but they are orthogonal to each other, that is, they are in quadrature. This means that $G_0(f)$ and $Q_0(f)$, or equivalently $g_0(t)$ and $q_0(t)$, can carry independent information even though they overlay each other in both time and in the frequency spectrum. We will use this property later when describing the

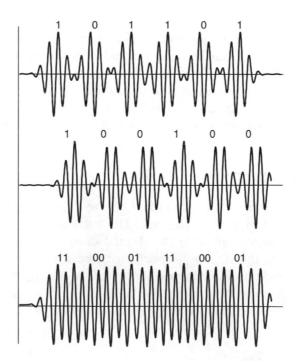

Figure 4.17 In-phase and quadrature signals in time.

modulation of signals to double the data capacity of a UWB signaling system in a given bandwidth.

4.6 Signal Modulation

Modulation is the process of modifying a signal wavelet, or impulse, so that it carries information. Here, we are interested in discrete digital representations of information. The simplest discrete states are binary states represented by "ones" and "minus ones". Any two distinct states can encode binary information. More than two states are often used, for example ternary (-1, 0, and $+1$). Generally speaking, we could have M states.

The modulation state of a UWB signal must be discerned in the presence of noise and interference to be recognized correctly. We must find one UWB signal among many other signals. Let us liken our desired UWB signal to a specific person, Kate, sitting in a large crowded auditorium shown in Figure 4.18. Kate signals by raising her right hand for a "one" and her left hand for a "minus one." We, the intended recipients of the message, know who Kate is; at least, we know that Kate is sitting in seat A11. We can ignore the rest of the motion among persons seated around Kate. This superfluous motion (see Figure 4.19) can be mistaken for a raised hand and is comparable to noise. We are looking for one of Kate's hands to be raised, and concentrate on Kate's sequence of right- and left-hand signals. Sometimes other signals may be mistaken for our desired signal. Henry, in seat A12 next to Kate, signals with his left hand, and perhaps blocks Kate's right-hand signal, as in Figure 4.20. This is signal interference. That signal might be interpreted as a "left" hand if

Figure 4.18   Kate signaling in a crowd [McKeown 2003].

Figure 4.19   A bustling crowd might be likened to a noisy signal environment [McKeown 2003].

Figure 4.20   Kate's hand signal is obscured by another signaler [McKeown 2003].

we mistakenly focus on Henry, whereas Kate actually signaled with her "right" hand. In that case, we would decode a signal "error." Messages must have enough redundancy so that we can reconstruct the missing parts, or mistaken parts of a message from those we have received correctly. One simple form of redundancy is simple repetition. We could ask Kate to send the same message more than once. There are redundancy techniques that are mathematically more refined than simple repetition and are referred to as Error Correction Codes.

Simple interference is not the only signal degradation. We might imagine Kate and Henry and many others in a house of mirrors. We see Kate and her many reflections, which now fill space and time. This kind of signal impairment is called multipath interference, as we will see when we investigate UWB signal propagation. Our key thought here is that the decoding of digitally modulated messages can result in errors. The error performance of modulations under specific conditions of noise and interference is called modulation efficiency. Modulation efficiency is a measure of the signal energy relative to the noise energy for a specified error rate.

We have described a binary signaling system: Kate raises her right hand for a "one" and her left hand for a "minus." We could additionally interpret a "zero" when she raises neither hand. That comprises three signal states or ternary modulation. Higher levels are possible: Kate could also signal by winking, nodding, and even shifting seats to represent M different signal states. These are called M-ary state, or level, modulations.

Developers of UWB impulse technology have perfected various ways for encoding information for transmissions. Pulses can be sent individually, in bursts, or in near-continuous streams, and they can encode information in pulse amplitude, polarity, and position. Modulations vary from simple pulse position, to the more energy efficient pulse polarity and to some very energy-efficient M-ary (multilevel or multistate) modulations.

The following commercially useful UWB impulse modulation techniques exemplify a wide range of implementation possibilities:

1. Pulse Position Modulation (PPM)

2. M-ary Bi-Orthogonal Keying (M-BOK) Modulation

3. Pulse Amplitude Modulation (PAM)

4. Transmitted Reference (TR)Modulation.

4.6.1 PPM Modulation

PPM modulation has been developed in the form of Time Modulation (TM) [Fullerton 1989] and was introduced by Time Domain Corporation in the late 1980s. It involves transmitting impulses at high rates, in the millions to tens of millions of impulses per second. However, the pulses are not necessarily evenly spaced in time, but rather they are spaced at random or pseudonoise (PN) time intervals as seen in Figure 4.21.

The process creates a noiselike signal in both the time and frequency domains. Time coding of the pulses allows for channelization, while the time dithering, fine pulse position, and signal polarity provide the modulation. UWB systems built around this technique and operating at very low RF power levels have demonstrated very impressive short- and long-range data links, positioning measurements accurate to within a few centimeters,

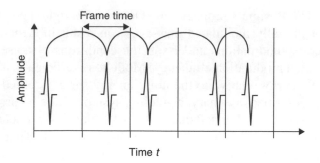

Figure 4.21 PN-coded UWB waveform sequence in time [Siwiak 2001].

and high-performance through-wall motion sensing radars. TM modulated systems use a fine pulse-shift modulation by positioning the pulse one-quarter of a signal cycle length (half the distance between two signal zero crossings) early or late relative to the nominal PN-coded location, or by pulse polarity. The error probability P_F of the fine-shift modulation in additive white Gaussian noise (AWGN) follows approximately the same behavior as conventional orthogonal or on-off keying (OOK),

$$P_F = \frac{1}{2} \operatorname{erfc} \left(\sqrt{\frac{\gamma_b}{2}} \right) \tag{4.13}$$

where γ_b is the received signal-to-noise ratio (SNR) per information bit.

4.6.2 M-ary Bi-Orthogonal Keying Modulation

Clever coding in combination with pulse polarity can result in a modulation efficiency that approaches the Shannon limit. One such modulation is *M-ary* Bi-Orthogonal Keying (*M*-BOK). This coding/modulation method is most easily applied in a DS-UWB system [McCorkle 2003]. A set of *M* moderate length (for example 24) ternary codes $(-1, 0, +1)$ is used to represent *M* symbols. Two codes in the set comprise 2-BOK modulation, which behaves like BPSK, while the 24-length chip sequence acts like a direct-sequence spreading code. When sent at 1,368 Mcps, 2-BOK enables 57 Mbps. With $M = 8$, 8-BOK allows 3 bits per symbol, and hence 114 Mbps. The *M*-BOK symbols are spaced on *M* orthogonal axes in the modulation symbol space, so the probability of symbol errors follows that of *M-ary* bi-orthogonal modulation [Proakis 1983]. Hence, in

AWGN the probability of a symbol error is

$$
P_m = 1 - \frac{1}{\sqrt{2\pi}} \int_{-\infty}^{\infty} \left(\frac{1}{\sqrt{2\pi}} \int_{-\infty}^{v+\sqrt{\frac{2\gamma M}{M-1}}} \exp\left(\frac{-x^2}{2} \right) \right)^{M-1} dx
$$

$$
\times \exp\left(\frac{-v^2}{2} \right) dv \tag{4.14}
$$

where $\gamma = \gamma_b \log(M)/\log(2)$ is the SNR per symbol and γ_b is the received SNR per information bit. The average bit error probability P_M is then

$$
P_M = P_m \frac{M}{2(M-1)} \tag{4.15}
$$

The *M-ary* orthogonal modulations like *M*-BOK have the property that as *M* becomes arbitrarily large the modulation efficiency approaches the Shannon limit. Said another way, when $M \rightarrow \infty$, the energy per bit to noise density ratio E_b/N_0 expressed in decibels approaches $-1.59\,\text{dB}$ for any bit error rate. That behavior is evident in Figure 4.22 where the error probability for various UWB modulations is shown as a function of the SNR per bit. The performance of *M*-BOK for $M = 2$ (same as BPSK), 4, 8, 16, and 64 are shown. *M*-BOK modulation may also be implemented using orthogonal impulse signals, for example, those of Equations (4.9) and (4.11). As with conventional QPSK, this serves to double the system capacity for the same occupied bandwidth.

4.6.3 Pulse Polarity, BPSK, and QPSK Modulation

Impulses sent individually with polarity modulation are analogous to BPSK in conventional carrier-based signaling. They perform the same as 2-BOK modulation shown in Figure 4.22. The error probability P_P of pulse polarity modulation follows the same behavior as conventional BPSK

$$
P_P = \tfrac{1}{2}\text{erfc}(\sqrt{\gamma_b}) \tag{4.16}
$$

where γ_b is the SNR per information bit. Equations (4.15) and (4.16) reduce exactly to Equation (4.16) for 2-BOK modulation. Impulses may be designed so that they are orthogonal as $g_0(t)$ and $q_0(t)$ in Equations (4.9) and (4.11). These may be superimposed in time

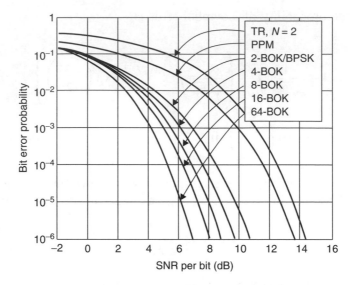

Figure 4.22   Probability of error for various UWB modulations.

and bandwidth to form the impulse equivalent of conventional QPSK modulation. The OFDM approach to a UWB system uses BPSK and QPSK modulation.

4.6.4 Pulse Amplitude Modulation

Impulses sent individually may be modulated by the pulse amplitude in an impulse variation of PAM. The PAM symbol error probability [Proakis 1983] is

$$P_a(\gamma_{\text{avg}}) = \frac{M-1}{M} \, \text{erfc} \, \left(\sqrt{\frac{3}{M^2-1} \gamma_{\text{avg}}} \right) \qquad (4.17)$$

for an average symbol SNR ratio. The corresponding bit error rate P_A as a function of average SNR per bit is

$$P_A = P_a \left(\gamma_b \frac{\log(M)}{\log(2)} \right) \frac{M}{(M-1)2} \qquad (4.18)$$

Figure 4.23 shows the bit error rate performance as a function of SNR per bit for $M = 2$, 4, 8, and 16.

Two-level PAM, 2-PAM, performance is equivalent to 2-BOK or BPSK; however, as M increases, the modulation becomes more fragile

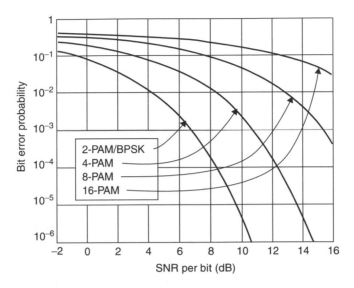

Figure 4.23 Probability of error for PAM modulation.

and requires higher SNR per bit. This modulation efficiency loss is approximately 4 dB for $M = 4$, and approaches 6 dB for every factor of 2 increase in M.

4.6.5 Transmitted Reference Modulation

Impulses can be sent with the information encoded differentially. A method of transmitting and receiving impulses that can easily implement a rake receiver is exemplified by a TR-UWB. The method employs differentially encoded *impulses* sent at a precise spacing D. The data value of the pulse is referenced to the polarity of the previously sent pulses. The system is shown in the simplified block diagram of Figure 4.24. The transmitter sends pulses separated by a delay D that are differentially encoded using pulse polarity. The pulses, including propagation induced multipath replicas, are received and detected using a self-correlator with one input fed directly and another input delayed by D. Long sequences of differentially encoded pulses may be sent in the same manner. The receiver resembles a conventional DPSK receiver, which, in AWGN, exhibits an error probability P_D, [Proakis 1983], of

$$P_D = \frac{1}{2} \exp\left(-\gamma_b \frac{N-1}{N}\right) \qquad (4.19)$$

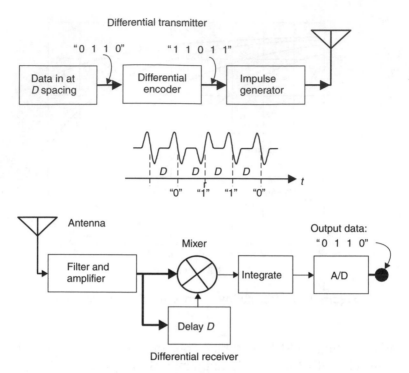

Figure 4.24   A TR-UWB transmitter and receiver.

where $N > 1$ is the number of differentially encoded pulses in a sequence and γ_b is the SNR per bit. The integration interval is sufficiently long to rake in a significant amount of the multipath energy. The TR-UWB receiver can rake in a significant amount of the multipath induced signal energy.

The SNR per bit performance in AWGN of the self-correlating TR receiver is shown in Figure 4.25 for $N = 2$ and for the limit of very large N. One novel channelization method employs pulse-pair TR-UWB (see [Hoctor 2001, Hoctor 2002]) with $N = 2$ and uses a family of delays D_i. Impulse pair sequences of these delay combinations comprise the channels. Figure 4.22 compares the error probability P_D of TR-UWB with $N = 2$ to the performance of other UWB modulations. Figure 4.25 compares P_D for $N = 2$ with that for N very large.

A more elaborate transmitter is shown in Figure 4.26. In comparison to the TR-UWB transmitter, this one produces precision shaped pulses and places them in precise locations in the spectrum. This transmitter is capable of generating precisely spaced pulses suitable for N-TR modulation where N is large. It can also produce M-BOK modulated pulse streams.

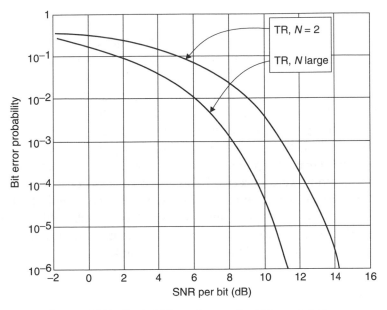

Figure 4.25   Probability of error for TR modulation.

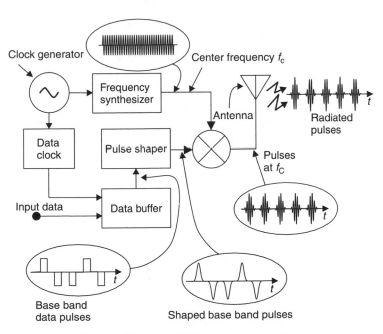

Figure 4.26   A UWB pulse transmitter.

4.7 Summary

In the chapter, we investigated how UWB signals might be generated under regulatory and physical constraints. The rules, such as those of the FCC, define access requirements to the spectrum rather than to UWB *technology*. Consequently, conventional radio technologies with parameters crafted to fit the rules will appear under the guise of "UWB." Impulse UWB signals may be generated by a great variety of methods, and several such methods were shown. Pulse shaping was found to determine the total pulse energy both within and outside the desired spectrum. Pulse modulations were described and the performance of various modulation techniques was compared. We saw that the simplest of UWB radios, a transmitted reference approach sported the least efficient modulation, while a pulsed direct-sequence UWB method employed M-BOK, the most energy efficient of the modulations studied.

References

[Aetherwire 2003] Aether Wire & Location, Inc., (Online): <http://www.aetherwire.com/>, 7 December 2003.

[Fullerton 1989] L. Fullerton, *Time Domain Transmission System*, U.S. Patent 4,813,057, 14 March 1989.

[Hoctor 2001] R. Hoctor, "Transmitted-reference, delay-hopped ultra-wideband communications", *Forum on Ultra-Wide Band*, Hillsboro, OR, (Online): <http://www.ieee.or.com/IEEEProgram-Committee/uwb/uwb.html>, 11–12 October 2001.

[Hoctor 2002] R. T. Hoctor and H. W. Tomlinson, *An Overview of Delay-Hopped, Transmitted-Reference RF Communications*, General Electric Company, 2001CRD198, January 2002, (Online): <http://www.crd.ge.com/cooltechnologies/pdf/2001crd198.pdf> 10 December 2003.

[IEEE802 03/157] IEEE802 Document: *Understanding UWB – Principles and Implications for Low-Power Communications – A Tutorial*, 10 March 2003, [03157r0P802-15_WG-Understanding_UWB_For_Low-Power_Commun.ppt].

[IEEE802 03/334] IEEE802 Document: [15-03-0334-05-003a-xtremespectrum-cfp-presentation.pdf].

[IEEE802 03/449] IEEE802 Document: [15-03-0449-03-003a-multi-band-ofdm-physical-layer-proposal-update.ppt].

[McCorkle 2003] J. McCorkle, "A comparison of DS-UWB and MB-OFDM techniques for high speed UWB networks", *UWB Summit '2003*, Paris, France, 2 December 2003.

[McKeown 2003]   D. McKeown, *Gammz UWB Cartoons and Art*, Private Communication to K. Siwiak, December 2003.

[Proakis 1983]   J. G. Proakis, *Digital Communications*, New York: McGraw-Hill, 1983.

[Siwiak 2001]   K. Siwiak and L. L. Huckabee, "Ultra wideband radio," in J. G. Proakis (Ed.), *Encyclopedia of Telecommunications*, New York: John Wiley & Sons, 2002.

5

Radiation of UWB Signals

Introduction

Short UWB pulse signals, when radiated, manifest some interesting differences compared to the behavior of narrowband signals. Short wideband signals have different shapes depending on where in the UWB radio link they are observed, while narrowband signals are sinusoidal everywhere. This is because the radiation of signals supplied to an antenna involves an acceleration of charges; and there is sometimes a partial time-delay derivative operation on the current density from the various parts on the antenna involved in the radiating process. There is no special magic in UWB propagation, no special exemption or dispensation from physical laws. UWB signals obey Maxwell's equations, but rather than considering them in frequency form, we prefer the time formulation. This is because instead of steady state harmonic wave solutions, we are interested in the transient responses. We need to abandon the narrowband (sine wave) simplifications to Maxwell's equations. We will see that for the antennas of maximum interest to us, the radiated UWB signals are related to the transmitter signal by a partial time-delay derivative, and that the received UWB signal is proportional to a weighted sum of time-delayed copies of the radiated signals. We will also see that because of the way UWB emissions regulations are crafted, the choice of the receiving antenna – whether it is a "constant gain" design like a dipole, or a "constant aperture" design like some horn antennas, or parabolic reflectors – can dramatically affect a UWB link performance. Specifically, if a "constant aperture" antenna can be employed on the receiving side of a UWB link, all frequencies then become of equal value. Additionally, we

Ultra-Wideband Radio Technology   Kazimierz Siwiak and Debra McKeown
© 2004 John Wiley & Sons, Ltd   ISBN: 0-470-85931-8

will see that certain antenna designs, which are equivalent for sinusoidal signals, can behave very differently for short pulses.

5.1 Short Pulse Radiation Process

Solving for the radiated fields of antennas involves finding solutions to Maxwell's equations in space and time. The equations are manipulated to account for boundary conditions, and finally placed into a second-order differential form called the *wave equation*. Inevitably, approximations and simplifications are employed to make the problem tractable. A convenient way of looking at many narrowband problems is to consider one frequency at a time; so, one of the most popular simplifications is to assume a sinusoidal solution for the wave equation. This involves using the time-honored time-harmonic function $\exp(j\omega t)$, which equals $\cos(\omega t) + j\sin(\omega t)$, in frequency ω. The time partial derivative operator $\partial/\partial t$ in Maxwell's equations then simplifies to a $j\omega$ multiplicative operator, and all the excess $\exp(j\omega t)$ terms are promptly suppressed and forgotten. After all, we know the answer is sine and cosine waves! We never worry about the time dependency of the signal at any point in the link. It is always a time-delayed sine function. Of course, the time-harmonic solutions (sine waves) are not the *only* solutions to Maxwell's equations (see [Harmuth 1968]); they are just the most commonly used and extensively applied solutions to a smorgasbord of narrowband and moderately wide systems, but not always useful for UWB systems.

Here, on the other hand, we are concerned with UWB solutions – solutions that involve wide bandwidth functions. Some derivations will be omitted or relegated to Appendix C, but results will focus on the differences between traditional time-harmonic solutions and those more appropriate for UWB signals. We will investigate the radiated fields due to arbitrary size dipoles to obtain general results, and then simplify to obtain results for the ideal infinitesimal antenna. Theoretical results will be compared with measurements. Simulations using analytical approximations will be compared with finite difference time domain (FDTD) methods.

We take a heuristic approach here with the goal of understanding the UWB radiation process and how it differs from the time-harmonic radiation solution. Although UWB under FCC regulations may include bandwidths as narrow as 5%, we consider here the short-time signal problem: UWB signals with bandwidths greater than about 40%.

Moving charges are called *electric currents*. Radiation in free space occurs when the current velocity changes, that is, the charges are *accelerated*. Charge acceleration, or the rate of change of charge velocity, is an electric current. When this current encounters certain discontinuities in conductors, radiation manifests itself as a partial derivative in time and delay of the current. Therefore, we expect that the *shape* of the transmitting antenna current signal plotted versus time will, in general, be different for the radiated signal. The difference between wideband and narrowband signals is that the transmitted signals in narrowband systems are sine waves, so their time derivatives are also sine waves. The signal *shape* for narrowband signals remains sinusoidal throughout the radiation process. The wider the signal bandwidth, the more a signal shape differs from its time derivative in amplitude versus time. We use the space time integral equation (STIE) technique (see [Bennett 1978]) to investigate the radiation problem analytically. This time-domain approach to electromagnetics allows us to study the short pulse radiation, which gives rise to ultrawide signal bandwidths. It lets us see the effects of time delays as the signal current densities interact at various points on the antennas.

We begin by defining the geometry of an arbitrarily shaped dipole antenna that supports surface currents $J(r', \tau)$ at points described by a vector r' pointing from the coordinate origin to the current density point, and by a retarded time variable $\tau = t - R/c$. The solution to the dipole current density on an arbitrarily shaped antenna is well beyond the scope of this work. We will obtain numerical solutions for a specific antenna shape using the FDTD numerical method described in [Kunz 1993]. The STIE formulation (see the geometry in Figure 5.1) allows us to express the radiated fields due to antenna currents in time–space rather than in frequency. Thus, transient or short pulse antenna excitations can be considered, rather than the steady state time-harmonic solution appropriate for narrowband techniques.

For our investigation of short pulse radiation, we choose a cosine modulated Gaussian pulse expressed by the equation

$$g_0(t) = \exp\left(\frac{-0.5t^2}{u_B^2}\right)\cos(2\pi f_C) \qquad (5.1)$$

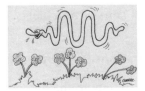

UWB Pulse in Free Space

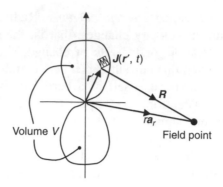

Figure 5.1   Dipole antenna with surface currents $J(r', \tau)$.

as the signal supplied to the terminals of the antennas. The choice of this signal is made for mathematical convenience, and there is no loss of generality in the following descriptions for other signal shapes. The 10-dB bandwidth of this pulse is $B = 2f_B$ and the parameter u_B is found from

$$u_B = \frac{1}{[2\pi f_B [\log(e)]^{1/2}]} \tag{5.2}$$

where $e = 2.81828\ldots$ is the base of the natural logarithms.

We relate the surface currents on the dipole to a feed-point current $I_T(t)$ supplied to a feed region of incremental length Δh and radius a; thus,

$$I_T(t) = \Delta h 2\pi a J(0, t) \tag{5.3a}$$

and $I_T(t)$ is proportional to signal $g_0(t)$ from Equation (5.1), so that

$$I_T(t) = (\Delta h 2\pi a) \exp\left[\frac{-(\pi t B)^2 \log(e)}{2}\right] \cos(2\pi f_C) \tag{5.3b}$$

and B is the signal 10-dB bandwidth. The current $I_T(t)$ is the signal supplied to the transmitting antenna, and as will be shown next, the radiated field strength is a weighted sum of time-delay derivatives of this current.

5.1.1 The Far-field of an Arbitrary Antenna

In Figure 5.1, the vector r' points to the antenna current vector J, and we are interested in the radiated field at point $r = ra_r$. An additional vector R

with magnitude R points from the current point to the field point. [Bennett 1978] gives the general expression for the magnetic far-zone field for the geometry in Figure 5.1 using the STIE formulation

$$H(r,t) = \frac{1}{4\pi rc} \int_V \frac{\partial}{\partial \tau} J(r', \tau) \times a_r \, dV' \qquad (5.4)$$

The integral over volume V reduces to an integration over the surface current density J, the *retarded time variable* is $\tau = t - R/c$, c is the speed of propagation, and a_r is a unit vector pointing in the direction of radiation. Equivalent expressions are given by [Baum 1989] for aperture antennas. The *retardation time R/c* is explicitly the propagation time required for a disturbance to travel the distance R at velocity c. The concept becomes obscured when sine wave solutions are assumed. For example, when the sinusoidal solution to the wave equation is given, the solution is in terms of $\cos(\omega\tau)$ or $\cos(\omega t - \omega R/c)$. The term $\omega R/c$ is then interpreted as the *phase* variable kR, thus losing the physical interpretation of *propagation delay* or retardation time. The electric far-zone field is

$$E = -\eta_0 a_r \times H \qquad (5.5)$$

and $\eta_0 = 376.73$ is the intrinsic free-space impedance. A key factor to note in Equation (5.4) is the appearance of the partial derivative in τ, which includes both time *and* delay. The magnetic field H and the electric field E are related to a time and delay derivative of the antenna current density; that is, to the acceleration of charges on the antenna. Because of this, the radiated field signal shapes will be different from the signal shape supplied to the feed point of the antenna. *This is a key difference from narrowband solutions where the signal shape everywhere is sinusoidal.* We should add that this time-delay derivative is not always present, as for example in some traveling wave antennas, which we will not consider here.

The fields calculated from currents in an antenna have been studied for over a century (see [Pocklington 1897, Hallén 1938]). These solutions are exceptionally difficult, even for simple wire antennas. We will not attempt such a solution here, nor do we need to for a heuristic understanding of the problem. Let us instead hypothesize a radiating structure such as the one pictured in Figure 5.2, in which a current is supplied to the feed point of a symmetrical wideband dipole. Radiation occurs from all parts of this antenna, as is represented by the integration in Equation (5.4),

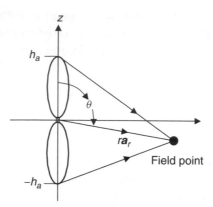

Figure 5.2   A wideband dipole.

but we will approximate this distributed source by a source at the feed
point and sources at each dipole end as shown in Figure 5.2. This three-
source model very closely resembles the exact expressions for radiation
from a thin wire dipole carrying a sinusoidal current distribution [Jordan
1968], and allows us to see the effects of time delay due to the finite
size of an antenna. Later, we will compare results using this simplified
approach to a complete FDTD numerical solution of this problem. Using
the simplified analysis, the approximate shape of the magnetic fields radiated from the dipole aligned with the z-axis and having effective length
$2h_a$ is

$$
H_\phi(r, t) = \frac{\sin(\theta)}{4\pi rc} \frac{h_a}{2} \frac{\partial}{\partial t} \left\{ I_z(t) + I_z \left(t - \frac{[1 - \cos(\theta)]h_a}{c} \right) \right.
$$
$$
\left. + I_z \left(t - \frac{[1 + \cos(\theta)]h_a}{c} \right) \right\} \tag{5.6}
$$

where $I_z(t)$ is the dipole feed-point current. The usual spherical (r, θ, ϕ)
coordinates are employed, and the $\sin(\theta)$ term is the projection of the
z-directed current density in the θ direction at the observation point.

The magnetic fields in Equation (5.6) appear to emanate from three
sources (see Figure 5.2), one at the feed point, and one at each dipole
end. The total magnetic field comprises time-derivative components that
are delayed by t as well as by $\cos(\theta)h_a/c$. Thus two processes contribute
to the magnetic field signal shape, a time delay and a distance delay.
*Both processes tend to lengthen the signal in time; hence both processes
will contribute to a narrowing of the radiated bandwidth compared to the*

bandwidth of the signal supplied to the antenna. Furthermore, and more generally, the delay term $\tau = t - \cos(\theta)R/c$ in Equations (5.4) and (5.6) means that the radiated signal shape as a function of time will vary depending on where the observation point is, measured by angle θ from the z-axis, relative to the antenna.

Figure 5.3 shows the transmitted signal, the antenna feed-point current, of Equation (5.1) using a 10-dB bandwidth of 4.5 GHz and center frequency $f_c = 4.5$ GHz, and the far-field signals along the plane of symmetry ($\theta = 90°$) as well as for $\theta = 45°$ and $30°$. Note that for this 100% wide bandwidth signal, the signal shape varies with angle θ. The lowest trace is the feed-point current. The next one above the current trace is the magnetic field strength in the broadside direction $\theta = 90°$ from the antenna. This signal shape most closely resembles a time derivative of the current waveform. As the observation angle θ moves towards the z-axis, the magnetic field waveform changes shape and its amplitude diminishes.

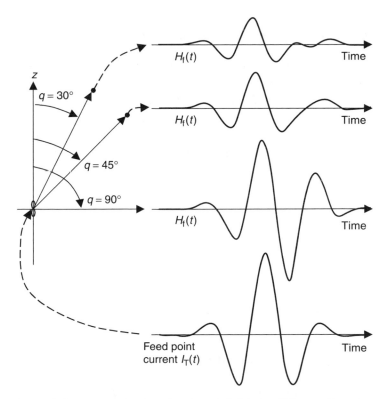

Figure 5.3 Transmitted and radiated fields in different directions.

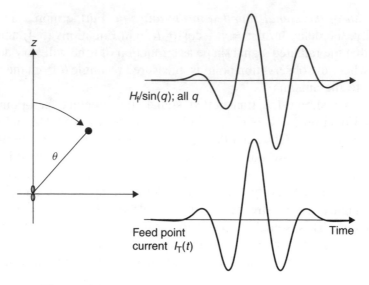

Figure 5.4   The infinitesimal transmitting antenna.

5.1.2 The Far-field of an Ideal Infinitesimal Radiator

The radiation picture can be simplified to its essence by reducing the antenna to an ideal infinitesimally small radiating current element. This radiating current element I has infinitesimal length Δh and radius a. In this case, the radiated magnetic field arises from just one of the sources in Equation (5.6), and we replace the surface current density by a z-directed feed-point current using the definition of Equation (5.2), so

$$H_\phi(r, t) = \frac{\Delta h \sin(\theta)}{4\pi rc} \frac{\partial}{\partial t} I_z(t) \qquad (5.7)$$

and the magnetic field is exactly a time derivative of the transmitting signal everywhere in space because the infinitesimal antenna is too small to impose any signal-distance delays on the antenna.

Figure 5.4 shows the geometry of the infinitesimal antenna along with the source signal and the radiated signal.

5.2 The Receiving Antenna

Antennas, for example dipoles, receive signals by a process of impressing incident fields onto the receiving antenna surface. These fields are weighted and summed up to appear as a voltage across the feed-point terminals of the receiving antenna. More specifically, reception involves impressing the incident electric field that would be present in the absence of the antenna, onto the conducting surface of the antenna. Then this incident field is projected onto the surface currents that would be present if the antenna were transmitting. The summation, or integral, of this process over the entire antenna is then scaled by the terminal feed-point current that would be present if the antenna were transmitting, resulting in the received signal voltage. The receiving process just described is a word statement of the *Principle of Reciprocity in Electromagnetic Theory* applied to the receiver antenna problem. Details and proofs are available in many textbooks including [Jordan 1968]. The reciprocity principle is a powerful tool for solving complex electromagnetics problems, and is applied here to arrive at the receiving properties of antennas for UWB signals. We care about the results here without the derivations, since we are interested in studying the effects of receiving short pulse signals. The effect of interest to us here is that the incident electric field undergoes a *time delay* before the fields impressed on the various parts of the antenna travel to the feed point to be summed up in the antenna load. There is no time derivative involved in receiving as there was in the transmitting process; however, the received voltage is not simply proportional to the incident electric field, but is rather a weighted summation of time-delayed points from the entire antenna surface. If the antenna were an infinitesimally small receiving antenna, then the time delays are zero and the received voltage is exactly proportional to the field strength.

5.2.1 The Arbitrarily Shaped Receiving Antenna

The signal received by an arbitrarily shaped dipole antenna is found by applying the principle of electromagnetic reciprocity, noting that the characteristics of an antenna while it is transmitting are the same as while it is receiving. The principle allows us to write an expression for the voltage signal at the receiving terminal of the antenna in terms of

1. the current density $J(r', \tau)$ that would be present if the *receive* antenna were used to transmit a feed-point current I_T;

2. the electric field E^{inc} due to the *transmit* antenna that would be incident on the receive-antenna location without the receive antenna being present.

Figure 5.5 shows the relationships between these various terms. From the reciprocity principle, the open circuit voltage at the receiving antenna terminals is

$$V_R(t) = \frac{-1}{I_T(t)} \int_V \boldsymbol{E}^{\text{inc}} \cdot \boldsymbol{J}(\boldsymbol{r'}, t) \, dV' \qquad (5.8)$$

The integration is over the current densities on the surface of volume V.

From Figure 5.5 it is clear that the received voltage signal V_R is related to the transmitted current signal I_T through several processes that can alter the shape of the signal. First, I_T gives rise to a current density distribution $\boldsymbol{J}$ on the transmitting antenna. This current density has time delays associated with it depending on the transmitter antenna dimensions. The current densities produce the electric field $\boldsymbol{E}^{\text{inc}}$ at the field point on the receiving antenna that is proportional to the time-delay derivative of the transmitting antenna current density. Finally, the received voltage V_R is a summation of weighted electric-field points, and another set of signal time delays is encountered in this summation because of the physical dimensions of the receiving antenna. A reasonable approximation for the voltage received by a wideband dipole is

$$V_r(t) = -h_a \left(E_z(t) + E_z \left(t - \frac{[1 - \cos(\theta)]h_a}{c} \right) \right.$$
$$\left. + E_z \left(t - \frac{[1 + \cos(\theta)]h_a}{c} \right) \right) \qquad (5.9)$$

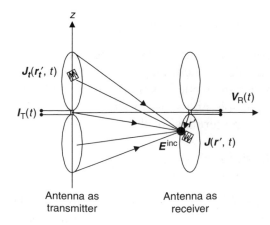

Figure 5.5   The receiving antenna.

Thus, the received signal V_R comprises weighted sum of and time-delayed derivatives of the transmitted current I_T. If the antenna is small, then there are no time-delay components in the received signal shape.

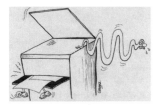

Delayed Copies

5.2.2 The Infinitesimal Receiving Antenna

When the receiving antenna is an infinitesimal "field probe" Δh incremental length oriented along the z-axis, and is $2\pi a$ in circumference, then the incident field is uniform across the antenna, and the integration in Equation (5.8) over the volume V containing current densities J reduces to $\Delta h I_T(t)$. There is no appreciable antenna length; hence there are no time delays with which to contend. The received voltage, therefore, is proportional to the incident electric field scaled by the incremental antenna length, as expressed by the equation

$$V_R(t) = -\Delta h E^{\text{inc}}(t) \qquad (5.10)$$

Equation (5.10) can also be deduced from the Lorentz Force Law, which relates the force F exerted on a charge q by the incident electric (E) and magnetic (H) fields on an antenna having length Δh.

$$F = Eq + q\mu_0 v \times H \qquad (5.11)$$

The antenna velocity v here is zero, so the magnetic field term drops out. The force per charge acting over the incremental antenna length has the units of voltage, so

$$V_R(t) = -\int_{-\Delta h/2}^{\Delta h/2} \frac{1}{q} F \cdot dl = -\int_{-\Delta h/2}^{\Delta h/2} E^{\text{inc}} \cdot dl = -E^{\text{inc}}(t)\Delta h \quad (5.12)$$

giving the same result as Equation (5.10).

If the transmitting antenna were also an infinitesimally small current element as shown in Figure 5.6, its transmitted electric fields, from Equations (5.5) and (5.7), would be proportional to the time derivative of the signal

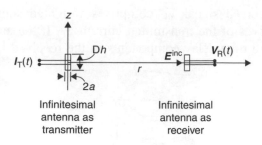

Figure 5.6   Infinitesimal transmitting and receiving antennas.

supplied to the antenna terminals. The received open circuit voltage at the terminals of an infinitesimal receiving antenna, then, for the pair of these antennas is the time derivative of the transmitted signal. For two "ideal" infinitesimal antennas placed a distance r apart and aligned for maximum signal,

$$V_R(t) = \frac{\eta_0 (\Delta h)^2}{4\pi r c} \frac{\partial}{\partial t} I_T(t) \qquad (5.13)$$

Equation (5.9) is also approximately correct for the shapes of signals transferred between two horn antennas aligned for maximum signal transfer (see [Baum 1989]). In that case the $(\Delta h)^2$ term is replaced by a term proportional to the horn antenna aperture area.

5.2.3 Transmission in Free Space Between Constant Gain Antennas

We can arrive at the free-space energy transmission formula starting with Equation (5.13) by realizing that the transmitted energy W_T (see Appendix C) is found by applying the transmitting current to the antenna radiation resistance of a Δh long infinitesimal dipole, which is given in [Siwiak 1998] as

$$R_{\text{rad}} = \frac{\eta_0}{6\pi} \left(\frac{\Delta h 2\pi f_c}{c} \right)^2 \qquad (5.14)$$

We also use the following approximation for the energy represented by the time derivative of the current

$$W_{\text{app}} = (2\pi f_c)^2 W_T \qquad (5.15)$$

With the details in Appendix C, the received energy W_R is found by applying the receiving antenna open circuit voltage $V_R(t)$ of Equation (5.10)

across a matched load equal to the antenna radiation resistance. Thus, the
energy transfer between infinitesimal dipoles is W_R / W_T. Finally, after some
algebraic manipulation, the free-space propagation equation for wideband
pulses is

$$P_L = \left(\frac{c}{4\pi f_c r} \right)^2 \tag{5.16}$$

We can now determine signal shapes and amplitudes anywhere in the link
between two antennas by using Equation (5.3b) for the transmit current,
with (5.15) for the transmitted energy. The field amplitude is found from
Equation (5.7), and the received voltage signal is given by (5.10), while
received energy is found by impressing half of the open circuit voltage
across a matched load resistance given by (5.14). Finally, a form of the
Friis transmission formula (see [Friis 1946]) for energy transfer between
two unity gain antennas in free space emerges as Equation (5.16).

5.2.4 Transmission with a Constant Aperture Receiving Antenna

The propagation process in free space is independent of frequency. Given
a fixed emitted energy, the energy density simply expands geometrically
and is collected by a receiving aperture. We saw in the previous section
that if the receiving aperture is a constant gain antenna, its receiving
aperture diminishes as the square of frequency imposing the frequency-
dependent behavior seen in Equation (5.16). This, of course, makes the
lower frequencies in the UWB frequency allocation more valuable for
radio links involving dipoles or other constant gain antennas. If, however, the receiving side of the link uses a constant aperture type of
antenna, like a pyramidal horn or a parabolic reflector antenna, the link
becomes independent of the frequency. With the details in Appendix C,
the propagation equation relating the radiated energy W_{EIRP} with the
energy W_R received by a horn antenna having aperture dimensions h_H
by h_W is

$$W_R / W_{\mathrm{EIRP}} = h_H h_W / \pi^3 r^2 \tag{5.17}$$

There is no frequency dependency! This sort of behavior is, naturally,
not unique to UWB emissions, but is general for all radio emissions.
In UWB, it means that *all frequencies are of equal value if a constant
aperture can be employed on the receiver side of a UWB transmission
link.*

5.3 Transmitted, Radiated, and Received Signals

In this section, we will see measured UWB signals compared with both FDTD and simplified analyses. The measurements were carried out on a 2-GHz center frequency system having a 10-dB bandwidth of about 150%. The measurements and the FDTD analysis are reported in [Siwiak 2001].

5.3.1 Simulations Using Wideband Signals

Measurements and simulations of signals in a UWB link between two diamond-shaped dipoles oriented for maximum signal transfer were performed for the geometry shown in Figure 5.7. The broadband antenna, introduced by [Masters 1947], comprises two flat conducting radiating elements shaped like triangles and arranged in the same plane with their bases parallel. They are fed symmetrically at the base. Measurements and FDTD calculations are shown in Figures 5.8 through 5.11. Figure 5.8 shows the measured simulated impulse transmitter waveforms, which correspond to the measured and simulated received waveform shown in Figure 5.9. Figures 5.10 and 5.11 extend this comparison to a transmitted signal comprising a sine-modulated transmitter pulse. Again, there is a close correspondence between measurements and analysis. Figures 5.12 and 5.13 show a comparison of waveforms calculated using the simplified analysis and those using the FDTD method. The simplified analysis gives results that are nearly identical to the detailed FDTD calculations.

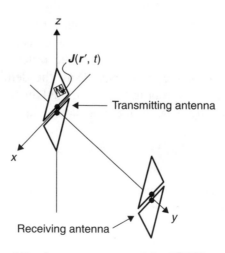

Figure 5.7   Antennas arranged for FDTD analysis.

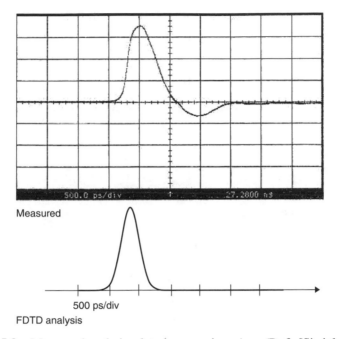

Figure 5.8   Measured and simulated source impulses (Ref: [Siwiak 2001]).

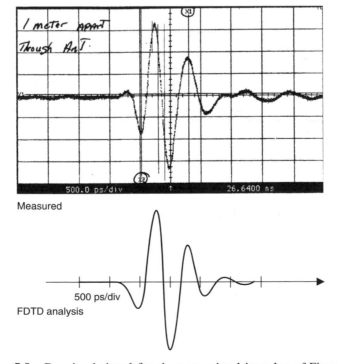

Figure 5.9   Received signal for the transmitted impulse of Figure 5.8.

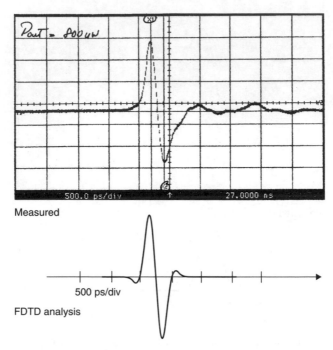

Figure 5.10   Measured and simulated source signals (Ref: [Siwiak 2001]).

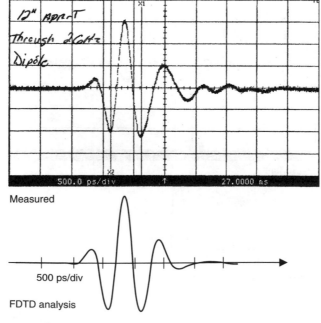

Figure 5.11   Received signal for the transmitted impulse of Figure 5.10.

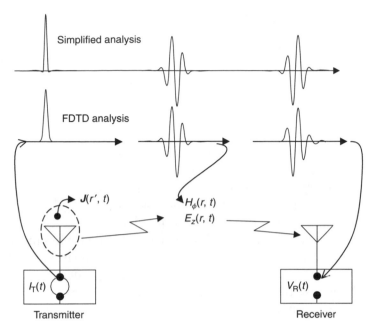

Figure 5.12 Source, fields, and received signals using FDTD and simplified analysis.

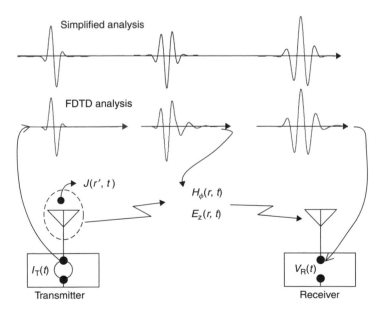

Figure 5.13 Source, fields, and received signals using FDTD and simplified analysis.

5.3.2 UWB at Moderate Bandwidths

Signal shapes have been calculated for a UWB link between two dipoles at a center frequency of 4.5 GHz and with signal 10-dB bandwidth of 2 GHz. The fractional bandwidth here is 44%. The signal shapes are shown in Figure 5.14. The transmitted signal is proportional to the current supplied to the transmitting antenna and has even symmetry around the point marked "A" in the figure. The radiated fields are expected to be related to the transmitted signal by a time derivative of a superposition of the transmitted signal and its time-delayed copies. As expected, this signal has odd symmetry around point "B" in Figure 5.14.

The received signal is a weighted composite of time-delayed copies of the radiated field signal arriving at the terminals of the receive antenna from various points on the transmit antenna, and likewise has the expected odd symmetry around point "C." The time alignment in Figure 5.14 between these three signals omits the direct radiation delay, so the slight additional time shift in the symmetry points "B" and "C" relative to "A" is due to the time delays arising from currents flowing on the finite

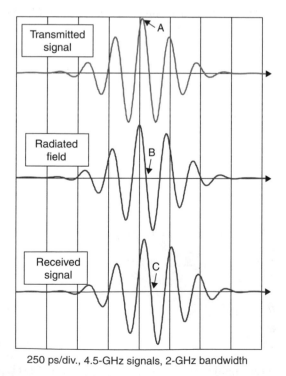

250 ps/div., 4.5-GHz signals, 2-GHz bandwidth

Figure 5.14   Signals in a 44% bandwidth UWB system.

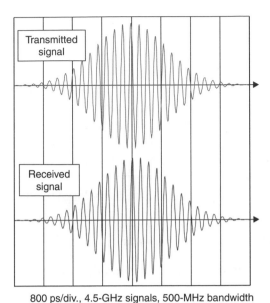

800 ps/div., 4.5-GHz signals, 500-MHz bandwidth

Figure 5.15   Signals in a 11% bandwidth UWB system.

sizes of the transmitting and receiving antennas. This kind of time-shift error would need to be taken into account when designing ultraprecise positioning measurements based on waveform symmetry points.

Figure 5.15 shows the transmitted and received signals for a 11% bandwidth system. The odd and even symmetries in the transmitted and received signals are present, but are no longer clearly discernible. A 500-MHz bandwidth pulse at 10 GHz would have more than twice as many cycles under the Gaussian envelope compared to the signal shown in Figure 5.15, and its bandwidth would be just 4.8%. While it is somewhat a stretch of the imagination to call this kind of a signal "UWB," it is in fact UWB under the FCC regulatory definition, and such emissions are permitted. This is a clear case in which the regulatory definition and the traditional understanding of UWB differ, but again it is worthwhile recalling that the regulations do not *define* a technology, but rather they define the rules under which spectrum may be accessed.

5.4 Some Antenna Effects in UWB

Antennas that are apparently the same when measured using a sine wave swept over a large bandwidth might exhibit different behaviors for wideband

UWB pulses. Here, we will look at two directional antennas that appear
to perform "equally" over wide frequency ranges when using narrowband
signals. With impulse excitations, however, they perform very differently.
We will first look at a horn antenna with an aperture excited in the TE10
mode. Then we will consider a pair of parabolic reflector antennas with
a wideband dipole at the focal points with dimensions chosen so that the
gain and impedance are essentially the same as the gain and impedance of
the horn antennas. We will then contrast their performance for sine wave
transmission and for pulse transmission. Finally, methods of broadbanding
antennas for multiband sine wave operation are not necessarily effective for
UWB signals, which require instantaneously large bandwidths.

5.4.1 The TE10 Mode Horn Antenna

A horn antenna operating in the TE10 mode has a gain given by

$$G_{\text{TE10}} = \frac{32 h_{\text{H}} h_{\text{W}} f_c^2}{\pi c^2} \tag{5.18}$$

As seen in Figure 5.16, a supplied transmitter pulse is propagated as
a partial time derivative of the sent pulse. In this simple illustrative
analysis, the effects of antenna currents on the edges of the horn aperture are ignored. In actuality, this effect can cause significant pulse distortion. The received pulse is then the same shape as the propagated

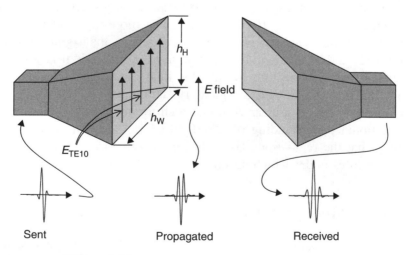

Figure 5.16   TE10-mode horn antennas in a link.

field pulse, but with inverted polarity. The pulse-field energy density and received signal energy are all adjusted for propagation attenuation and for antenna gains.

If a sine wave had been supplied to the transmitting horn, the propagated field would be a sine wave and the received signal would be a sine wave. All the sine wave amplitudes are adjusted for propagation attenuation and antenna gains. In this case, with the exception of pulse shape, the antennas transmit pulses the same way as they do sine waves.

5.4.2 The Dipole-fed Parabolic Reflector Antenna

Let us now look at parabolic reflector antennas fed by wideband dipoles at their focal points. The parabola dimensions are chosen so that the antenna gains equal the horn antenna gains in Figure 5.16. The wideband dipoles, similarly, are designed so that their impedance bandwidth match those of the horn antennas. The link between two such reflector antennas is shown in cross-section in Figure 5.17. A pulse sent by the transmitter is propagated by the feed-point dipole towards the reflector and also in the transmission direction. The result is a propagated field-strength profile that includes a leading pulse directly radiated by the dipole and a pulse $2a$ behind it that reflects from the parabolic surface and picks up the antenna

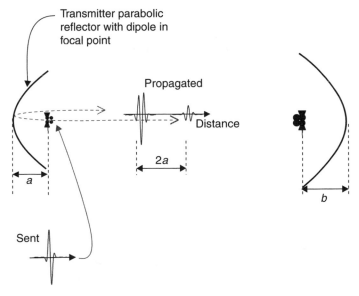

Figure 5.17 Dipole-fed parabolic reflector antennas in a link.

area aperture gain. The reflected pulse polarity is inverted relative to the directly radiated pulse.

The pair of pulses is now incident on a receiving parabolic reflector antenna shown in cross-section in Figure 5.18. The dipole at the focus of the antenna receives the pair of directly propagated pulses, and also receives the same pair of pulses reflected from the receiving parabola, now delayed by the added distance traveled and magnified by its gain aperture. Thus four pulses are received. Two are received directly by the receiving dipole and two more are delayed by the additional path to the receiving dipole via the receiving parabolic reflector. The received pulse polarities include a pulse inversion for each reflection in the total transmission path as well as an inversion due to the receiving process. One of the received pulses does not benefit at all from any reflector aperture gain. Another two pulses each benefit from one reflector magnification, and a single pulse benefits from the aperture gain of both parabolic reflectors. Once again, in this illustrative example, effects due to signal interaction with the edges of the parabolas are ignored.

For pulse transmission, one pulse is sent, and up to four pulses are received. When the same transmission link is used for a sine wave signal, the receive signal is a summation of four time-delayed sine waves of four different amplitudes and phases. That summation is a single sine wave. For sine wave transmissions, the systems of Figures 5.16 and 5.18 appear identical. For pulses, they are demonstrably different.

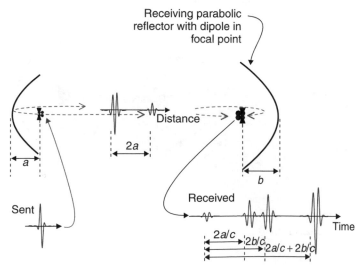

Figure 5.18 Link between two dipole-fed parabolic reflector antennas.

The particular antenna system of Figure 5.18 effectively generates a form of multipath transmission for pulses.

5.4.3 Wideband Antenna Considerations

We have seen a few examples of how short impulses can behave differently from sinusoidal signals in the antenna radiation process. The topic of wideband antennas suitable for UWB impulses deserves a textbook-length treatment, and only a few short survey comments are included here. Antennas must be understood in their time-domain behavior rather than in the frequency domain, because the effects of short pulses become observable whenever the propagation delay, the *retardation time*, on the antenna is similar to the impulse rise time or duration. For short pulses, only certain segments of the antenna are involved in the radiation process at any one time. Wideband antennas have typically been designed with a specification of flat amplitude and linear phase response over the desired bandwidth (a frequency-domain approach) without regard to out-of-band performance. This approach does not guarantee an adequate design with respect to "settling time" or transient response in an antenna. The result presents residual time-domain artifacts: ringing responses in the antenna. The out-of-band antenna response significantly shapes the total time-domain response of the antenna. For sine wave excitations, the antenna response is, after several RF cycles, equal to the sinusoidal forcing function – sine wave in and sine wave out – across the entire operating bandwidth of the antenna. For short impulse excitations, the antenna output is a transient response, which includes all of the natural resonant responses of the structure with amplitudes that may be commensurate with the forced impulse response. Resonant effects were not factors in the carefully chosen examples of the previous sections.

5.5 Summary

The radiation of UWB signals involves fields that are time-delayed time derivatives of the signal currents from the various parts of the transmitting antenna. For antennas that resemble an infinitesimal current element, there are no time delays on the antenna and the radiated far-fields are shaped exactly like the time derivative of the signal applied to the radiating current element. The reception process at the receiving antenna involves a time-delayed integration of the radiated signal fields over the receiving antenna surface. Again, for an infinitesimal current element antenna, there are no time delays on the antenna, and the received signal is proportional to the signal field strength. Thus, the transmitting signal, the field signal, and the received signal differ in shape. A simplified analysis was developed and was verified by comparisons with measurements and with FDTD simulations. We noted that dipoles are representative of constant gain antennas, in contrast to constant aperture antennas, which, when used as receive antennas, render the propagation link independent of frequency. We also saw that gain antennas having essentially equal performance for sine waves, even over large bandwidths, can perform very differently for pulses, introducing a form of multipath transmission. Finally, we noted that UWB antenna designs need to be considered in the time domain, with due attention to their transient responses and to the time-delay nature of radiation from the various segments of the antenna.

References

[Baum 1989] C. E. Baum, "Radiation of impulse-like transient fields", *Sensor and Simulation Note 321 USAF Phillips Lab*, Albuquerque, NM, November 1989.

[Bennett 1978] C. L. Bennett and G. F. Ross, "Time-domain electromagnetics and its applications", *Proceedings of the IEEE*, **66**(3), 299–318. 1978.

[Friis 1946] H. T. Friis. "A note on a simple transmission formula", *Proceedings of the IRE*, **34**, 245–256; **41**, 1946.

[Hallén 1938] E. Hallén. "Theoretical investigation into transmitting and receiving qualities of antennae", *Nova Acta Regiae Societatis Scientiarum Upsaliensis*, **II**(4), 1–44, 1938.

[Harmuth 1968] H. F. Harmuth, "A general concept of frequency and some applications", *IEEE Transactions on Information Theory*, **IT14**(3), 375–381, 1968.

[Jordan 1968] E. C. Jordan and K. G. Balmain, *Electromagnetic Waves and Radiating Systems*, Second Edition, Englewood Cliffs, NJ: Prentice Hall, 1968.

[Kunz 1993] K. S. Kunz and R. J. Leubbers, *The Finite Difference Time Domain Method for Electromagnetics*, Boca Raton, FL: CRC Press, 1993.

[Masters 1947] R. W. Masters, *Antenna*, US Patent 2,430,350, November 4 1947.

[McKeown 2003] D. McKeown, *Gammz UWB Cartoons and Art*, Private Communication to K. Siwiak, December 2003.

[Pocklington 1897] H. C. Pocklington, "Electrical oscillations in wires", *Proceedings of the Cambridge Philosophical Society*, **9**, 324–333, 1897.

[Siwiak 1998] K. Siwiak, *Radiowave Propagation and Antennas for Personal Communications*, Second Edition, Norwood, MA: Artech House, 1998.

[Siwiak 2001] K. Siwiak, T. M. Babij and Z. Yang, "FDTD simulations of ultra-wideband impulse transmissions", *IEEE Radio and Wireless Conference: RAWCON2001*, <http://rawcon.org>, Boston, MA, August 19–22, 2001.

6

Propagation of UWB Signals

Introduction

In the previous chapter, we developed the radiation process for UWB signals in an unobstructed free-space path. Here, we examine real-world modifications to that path. The "free-space" propagation model is fully deterministic, and is exactly the same as for time-harmonic (sinusoidal, or narrowband) signals, except that care must be taken in defining the "center frequency" for the propagation. We will, additionally, consider reflections from and transmissions through materials. Energy expands geometrically, so the energy density decreases with the square of distance. The simplest case involving reflections is a transmission link above a planar surface, which is the "two-path" model.

The two-path model is fully deterministic. It depends on the exact shape and frequency of the UWB signal because there is a transition region between free-space behavior and 4th power law behavior. The transition region is where the direct and reflected signals partially overlap, resulting in an interference pattern dependent on the specific shape and characteristics of the UWB signal. This model is applicable to propagation between antennas that are generally in the clear, and the cases in which signal paths can be accurately described by an unobstructed direct path plus a ground-reflected path.

When propagating among the clutter of an indoor environment (see [Foerster 2001, Siwiak 2001, Siwiak 2001a]), UWB impulses shed energy because of reflections that manifest themselves as time-dispersed echoes. This shedding of energy is, in addition to the energy density decrease because of geometrical spreading. There is no "fading" in the classic time-harmonic signal sense because UWB signals are short in time and

Ultra-Wideband Radio Technology   Kazimierz Siwiak and Debra McKeown
© 2004 John Wiley & Sons, Ltd   ISBN: 0-470-85931-8

space. Multipath fading, a characteristic of conventional (time-harmonic) RF communications, is the result of coherent interaction of sinusoidal signals arriving by many paths. Some communications channels, particularly outdoors, can have multipath dispersion that is measured in many microseconds; therefore, some multipath components can be resolved and received using rake techniques. In buildings, however, the multipath differential delays are in the several tens of nanoseconds and cannot be resolved in the relatively narrow conventional radio channels. These systems must, therefore, contend with significant Rayleigh fading, which may require signals up to tens of decibels above the static signal level for a given measure of performance. UWB signals are commensurate with indoor signal dispersion, and can resolve the multipath. There is no Rayleigh fading but rather dispersion of the signals in time.

An indoor statistics-based multipath model will be developed here, which provides a description of a median behavior of signals in full multipath. The model is novel in that it provides a theoretical basis for propagation laws other than the square law by accounting for the energy shed by signals to multipath reflections. The theory also predicts maximum possible rake gain based on collecting the time-dispersed reflections. This model provides a smooth transition between free-space behavior and another propagation power law that takes effect beyond some breakpoint distance. With the proper choice of parameters, it can be applied as a simplified model for two signal paths.

6.1 Signal Propagation in Free Space

UWB signals propagate in free space as three-dimensional geometrically expanding waves. Beyond the immediate near-field vicinity of the antenna, propagating energy expands spherically in proportion to the square of the distance. For omnidirectional expansion, the total energy remains constant over the surface area $4\pi d^2$ of a sphere having a radius d. *Hence, the propagation of signal energy density is independent of the frequency.* The energy density propagates as $1/4\pi d^2$ as shown in Figure 6.1.

Energy is then captured by a unity-gain antenna having capture area $A_e = \lambda^2/4\pi$, where the wavelength $\lambda = c/f_m$ is written in terms of frequency f_m and the speed of propagation c. We obtain a form of the Friis transmission formula (see [Friis 1946])

$$P_L = 20\log\left(\frac{c}{4\pi d f_m}\right) \tag{6.1}$$

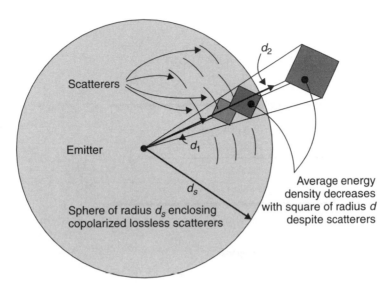

Figure 6.1   Spherical spreading of a signal.

describing energy transfer between the terminals of two unity-gain antennas. This formula is a product of a frequency-independent wave expansion term and the frequency-dependent antenna capture–area term. The frequency dependency of propagation comes from the frequency dependency of the "constant gain" receiving antenna aperture area $A_e = c^2/4\pi f^2$ evaluated here at f_m. The total received power involves an integral in frequency over the product of the received power spectral density and A_e. This integration approximately results in a frequency f_m, which is the geometrical mean of the upper and lower signal frequency limits.

6.2 Propagation with a Ground Reflection

Propagation over a smooth surface involves a reflection from the ground as shown in Figures 6.2a, 6.2b and 6.2c. The direct-path and reflected-path lengths D and R in terms of the antenna heights H_1 and H_2, and the separation distance d are derived from simple geometry

$$D = \sqrt{d^2 + (H_1 - H_2)^2} \tag{6.2}$$

and

$$R = \sqrt{d^2 + (H_1 + H_2)^2} \tag{6.3}$$

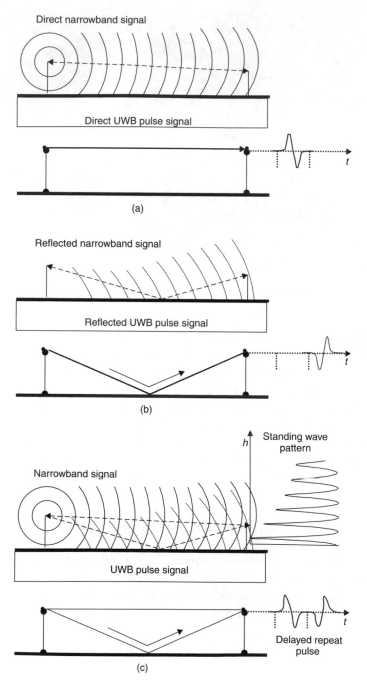

Figure 6.2 (a) Narrowband and UWB direct-signal path; (b) narrowband and UWB reflected signal path; (c) direct and reflected signals in two-path propagation.

(see [Siwiak 1998] for the details). The differential delay between the reflected path and the direct path over a plane earth is

$$\Delta t = \frac{(R - D)}{c} \tag{6.4}$$

where c is the speed of propagation.

Figure 6.2a shows propagation of the signal by the direct path for both the narrowband signal and a UWB pulse signal. The narrowband signal, approximated by a sine wave, remains on continuously. The pulse is a transient phenomenon and is present only for its short duration. Figure 6.2b shows the signals following a path that reflects from the ground. The narrowband signal is again a sine wave, so it persists forever, while the reflected impulse is simply a delayed polarity-inverted copy of the direct impulse. Figure 6.2c shows the composite direct and reflected signals. The direct-path sine wave and the reflected-path sine wave both persist forever. As a result, they form a vertical standing wave pattern whose magnitude is depicted on the right-hand side of Figure 6.2c. The UWB propagation is a transient phenomenon. At the receiver end, the direct-path pulse arrives first, followed by the reflected-path pulse. Unless the difference between the direct and reflected paths is shorter than the pulse length, the pulses will be received distinctly and will not overlap in time.

6.2.1 UWB and Time-harmonic Signals with a Ground Reflection

The ground reflection coefficient for the cases of interest (shallow incidence angles) is very nearly -1, so the reflected signal undergoes a simple polarity inversion. Reflected signals that arrive by paths having differential delays greater than the signal duration do not overlap in space–time, as portrayed in Figure 6.3, and can add to the total received energy if pulse-rake techniques are employed.

Shallow Angle of Incidence

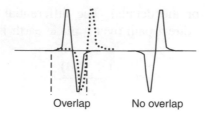

Overlap   No overlap

Figure 6.3   Overlapping and nonoverlapping pulses.

Pulse signals with a differential delay of *less* than the pulse length begin to exhibit constructive and destructive interference in the receive window.

The overlapping signal pulse echo arriving by ground reflection is not delayed enough to be distinct from the directly arriving pulse. The general propagation behavior of the signal pulse propagated between two antennas over a smooth ground is pictured in Figure 6.4. In the nonoverlap region, the propagation is governed by the inverse square law given by Equation (6.1). With a rake receiver, energy from both direct and reflected signal impulses could be collected and the overall propagation represented by the bold line in Figure 6.4 in the "nonoverlap region" would be raised by 3 dB. Overlapping pulses can add constructively when the overlap is partial, then, when nearly fully overlapping, they can exhibit an inverse 4th power with distance behavior beyond the breakpoint distance d_t, similar to that of time-harmonic wave propagation (see, for example, the detailed derivation in [Siwiak 1998]).

It is easy to show that the two-path propagation law in the overlap region beyond the breakpoint distance d_t is inverse 4th power and given by

$$P_{2\text{path}} = 20\log\left(\frac{h_1 h_2}{d^2}\right) \tag{6.5}$$

This law holds for both UWB and time-harmonic signals. It is noteworthy that there is no frequency dependency in Equation (6.5). The breakpoint distance d_t between the inverse square law region governed by Equation (6.1) and the inverse 4th power law governed by Equation (6.5) is found by equating Equations (6.1) and (6.5) and solving for distance

$$d_t = \frac{4\pi f_m h_1 h_2}{c} \tag{6.6}$$

The breakpoint distance d_t clearly depends on the geometry, that is, on the antenna heights as well as on the frequency of the propagation.

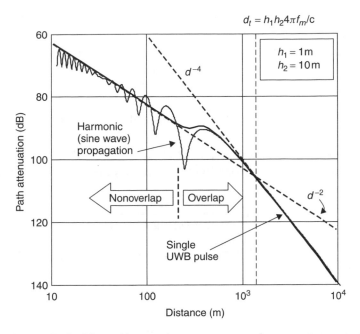

Figure 6.4 UWB (bold), and harmonic-wave propagation near the ground. Reproduced by permission of John Wiley & Sons, Inc.

Compared with UWB signals, harmonic waves exhibit multiple constructive and deep destructive interferences as multiple sinusoidal cycles interact at close distances. Both the sinusoidal (time-harmonic) and the short UWB signals behave identically beyond the breakpoint distance, as seen in Figure 6.4. The detailed behavior of UWB signals in the transition region between square law region and the breakpoint distance depends on the interference pattern between a pulse signal and its delayed polarity-inverted copy; hence, it depends on the details of the UWB signal structure.

6.2.2 Design Example of a 2-GHz UWB Wide Signal

In this section, we consider the propagation of a specific UWB signal: one having a 2-GHz bandwidth defined at the 10-dB points and with a center frequency of 4.5 GHz. The signal comprises a base band bell-shaped Gaussian pulse $g(t)$ described by the "base band" time signal, written for graphing convenience with its peak value set to 1:

$$g(t) = \exp\left(\frac{-0.5t^2}{u^2}\right) \tag{6.7}$$

where u is the pulse-width parameter. This "time" signal pulse can be equivalently written as a function of frequency, again with a peak value set to 1 for convenience, so we can view how it occupies the spectrum:

$$G(f) = \exp[-2(\pi f u)^2] \qquad (6.8)$$

Mathematically, the time function $g(t)$ and its frequency representation $G(f)$ are related by the Fourier transform, but we need not dwell on that detail here. We choose a particular $u = u_B$ so that $G(f)^2 = 0.1$ at $f_B = 1.0\,\text{GHz}$ to meet the bandwidth requirements as seen in Figure 6.5. Solving for u_B,

$$u_B = \frac{1}{[2\pi f_B[\log(e)]^{1/2}]} \qquad (6.9)$$

and f_B is the desired half bandwidth of 1 GHz and e $= 2.81828\ldots$ is the base of the natural logarithms.

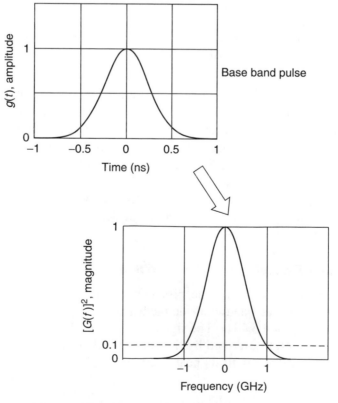

Figure 6.5   Base band signals $g(t)$ and $G(f)$.

The final step in this UWB signal design example is to frequency shift (heterodyne) the base band pulse $g(t)$ up to the desired, operating center frequency of 4.5 GHz. This comprises a simple mixing with or multiplication by a cosine wave centered at the desired center frequency of $f_C = 4.5$ GHz. The frequency-shifted signal is

$$g_0(t) = \exp\left(\frac{-0.5t^2}{u_B^2}\right)\cos(2\pi f_C) \qquad (6.10)$$

This appears in the frequency spectrum as

$$G_0(f) = \exp[-2(\pi[f \pm f_C]u_B)^2] \qquad (6.11a)$$

as seen in Figure 6.6. Signal $g_0(t)$ is not the signal supplied to the antenna, rather, it is the radiated signal, that is, it has already undergone the radiation process from an antenna. The characteristics of $G_0(f)$ meet our design criteria. The signal is 10 dB below the peak level at 3.5 GHz and is also 10 dB below the peak level at 5.5 GHz for a bandwidth of 2 GHz.

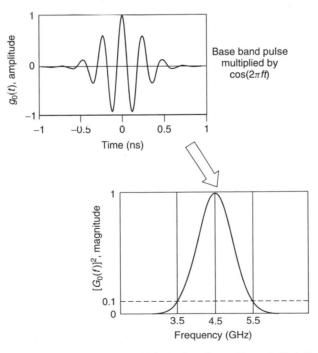

Figure 6.6 Frequency-shifted signals $g_0(t)$ and $G_0(f)$.

The current representing a pulse with 10-dB bandwidth B supplied to the terminals of an antenna may be written as

$$I_T(t) = I_0 \exp\left[\frac{-(\pi t B)^2 \log(e)}{2}\right] \cos(2\pi f_C) \qquad (6.11b)$$

and the total energy delivered to the antenna is then the time integral of the square of $I_T(t)$ multiplied by the antenna radiation resistance.

6.2.3 EIRP of the 2-GHz Bandwidth Pulse

A single isolated pulse delivers *energy*, measured in joules (J). Energy pulses are delivered at an average signal pulse repetition frequency (PRF), so they appear as energy per unit time (J/s), or power. The emission rules and regulations specify limits on effective isotropically radiated power (EIRP), that is, power radiated by an antenna having a gain of 1. Let us consider a repetitively sent signal $G_0(f)$ whose power spectral density (PSD) is centered at a frequency f_C, and whose peak amplitude has been set to 1 for convenience, as pictured in Figure 6.6. The total emitted power is then found by computing the area under the curve described by the PSD and multiplying it by $10^{-41.3/10}$ mW/MHz. This area calculation and multiplication, when expressed in decibels, give the total power in dBm:

$$P_{EIRP} = 10 \log\left(\int_0^\infty G_0(f)^2 df \; 10^{-41.3/10} 10^{-6}\right) \qquad (6.12a)$$

The frequency f is in hertz.

The closer that $G_0(f)$ conforms to the area under the PSD limits (as seen in Figure 6.7), the more power the signal can be made to transmit. If all available spectrums from 3.1 to 10.6 GHz were perfectly filled with the maximum allowed signal PSD, the total EIRP for a repetitively sent signal would amount to $(10,600-3,100) \; 10^{-41.3/10} = 0.556$ mW, equivalent to -2.55 dBm. This represents the absolute maximum possible EIRP limit for UWB under these particular regulations. The maximum allowable EIRP of this pulse is found by integrating the square of $G_0(f)$ over frequency, and multiplying the result by the peak PSD limit permitted under regulations. In practical designs, some margin must be allowed for spectral peaking due to any regularities in the modulation of the pulses. In this example, the power or the EIRP is $P_{EIRP} = 86\,\mu\text{W}(-10.6\,\text{dBm})$ for a pulse of this design sent at some PRF.

Table 6.1 lists the maximum possible EIRP as well as the characteristics of a 500-MHz bandwidth UWB pulse in comparison with the 2-GHz wide

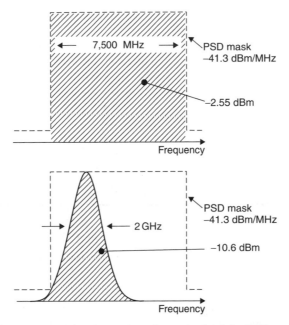

Figure 6.7 The pulse energy depends on how much of the PSD mask is occupied.

Table 6.1 UWB signal design characteristics for a Gaussian envelope pulse.

Pulse type	Bandwidth at the 10-dB point	Center frequency, f_C (GHz)	Maximum EIRP (dBm)
na	7.5 GHz	na	−2.55
Gaussian	2 GHz	4.5	−10.6
Gaussian	500 MHz	3.46–10.24 GHz	−16.6

design example pulse. This is the smallest bandwidth permitted under the FCC rules governing UWB.

6.2.4 Propagation of a 2-GHz-Wide UWB Signal Near the Ground

The signal $g_0(t)$ takes two paths between the transmitting and receiving antennas shown in Figure 6.2c. There is a direct path, and a reflected path where the reflected signal undergoes a polarity change. The differential distance between the two paths decreases as the range d between the antennas increases. Thus, the direct pulse and the reflected pulse arrive at the receiver antenna closer and closer together as the distance increases.

Table 6.2 Direct and reflected signal relationships between antennas 1 m above a ground.

Range: m	Differential distance: m	Differential time: Δt (ns)	Energy relative to free space: W (dB)
2	0.83	2.8	3
4	0.47	1.6	3
10	0.20	0.66	2.3
30	0.067	0.22	−4.2
50	0.040	0.13	5.4
300	0.007	0.02	−14

Table 6.2 shows the differential distances and the differential time of arrival for various distances for this geometry with antennas 1 m above a ground.

At a range of 2 m, the reflected path distance is 0.83 m longer than the direct path, so the direct signal and reflected signal are displaced $0.83/c = 2.8$ ns. Also, there are two pulses available, so a rake receiver can gather an additional 3 dB of signal relative to free-space propagation. The relative energy reported in Table 6.2 is found by integrating the square of the sum of the direct and polarity-inverted reflected signal:

$$W = 10 \log \left(\frac{1}{W_0} \int_{-\infty}^{\infty} [g_0(t) - g_0(t + \Delta t)]^2 \right) dt \qquad (6.12\text{b})$$

where W_0 is the energy of the single pulse. Since two pulses are available in the nonoverlapping region close to the antenna, the energy there is 3 dB above the free-space value.

Figure 6.8 shows the direct and ground-reflected impulses for UWB propagation between two antennas 1 m above a ground for different separation distances between the two antennas. When the two antennas are close together, the differential path distance is greatest, so the time difference of arrival between the direct and reflected signals is greatest. As the separation distance between the two antennas is increased, the time difference of arrival between the direct and reflected signals becomes smaller. Eventually, the signals are close enough together to overlap and add constructively and destructively, depending on the distance.

Figure 6.9 shows the total signal amplitude plotted versus distance with these same signals identified with particular distances. The effect on the total signal amplitude of the overlapping signals along with the constructive and destructive interference is clearly evident.

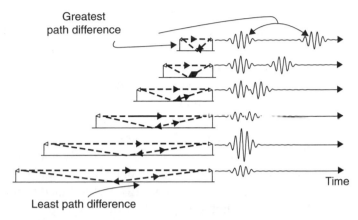

Figure 6.8 UWB signals propagating near the ground.

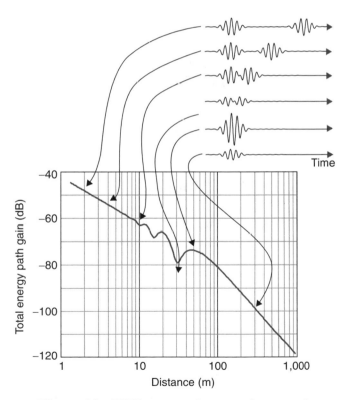

Figure 6.9 UWB propagation near the ground.

The signal relationships are depicted pictorially in Figure 6.9. For this geometry, the breakpoint distance $d_t = 183.9$ m. Beyond this distance, the signal energy diminishes at a rate of $1/d^2$ on top of the $1/d^2$ rate because of geometrical expansion of the wave for a net propagation law of $1/d^4$. Figure 6.9 shows the propagation performance of a UWB link between two unity-gain antennas that are each 1 m above the ground, that is, $h_1 = h_2 = 1$ m, and separated by distance d. The UWB signals arriving at the receiving antenna are pictured for several distances. Both Table 6.2 and Figure 6.9 show that there is a 4.2-dB destructive interference or "null" at the 30-m range for this choice of antenna heights and pulse-center frequency. On the other hand, there is a 5.4-dB performance improvement due to constructive interference compared to the single-pulse free-space performance at a 50-m distance. Other pulse shapes and other choices in antenna heights will give different results.

Breakpoint Distance

6.3 Propagation of UWB Impulses in Multipath

A short UWB impulse traversing the inside of a building may encounter walls and obstacles, which provide many opportunities for creating signal echoes. These echoes are evident as multiple delayed reflections of the first-arriving impulse adding to the signal amplitude in time. "Remember 'Kate in the crowd in Figures 4.18–4.20 when we discussed modulation? *Multipath is comparable to 'Kate in a House of Mirrors'*." Recent high-resolution UWB channel impulse response (CIR) measurements reveal different path-loss exponents in the distance dependence of the small area average for the total received energy and the energy of the strongest arriving pulse [IEEE802 02/301]. It is also found that the signal delay spread, the time record of echoes, increases with distance. This leads us to a simple theory, reported in [Siwiak 2003], which connects the differences in the propagation exponent with multipath scattering. This result has implications on receiving signals in multipath, rake gain, and understanding the multipath wireless channel.

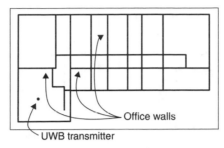

Figure 6.10   Office area with a UWB source.

6.3.1 An Impulse Propagating through a Building

UWB propagation is a transient phenomenon. A pulse, unlike a sine wave, has a very short duration in both time and space. Figure 6.10 shows a top view of a simulated office area having a UWB transmitter in the lower left corner, which emits a single UWB pulse.

The small-width pulse expands spherically like an expanding shell from this transmission point, as seen in Figure 6.11, like a wave in a pond or a pool of water [Lauer 1994, Zollinger 2002]. There is no energy in front of the shell, and initially no energy following the expanding shell.

In Figure 6.12, the wave has already expanded beyond the immediate room and has traversed the first walls and continues to expand through them. Reflections of the initial wave front are now evident as waves traveling back to the source point.

Further wave-front expansion can be seen in Figure 6.13 where the wave front has traversed several walls, and multiple echoes and reflections become evident. The multipath, of course, is confined to the area within the leading edge of the expanding wave front.

In Figure 6.14, the wave front has expanded to most of the office area. Only echoes remain behind the wave front. An observer at any point in

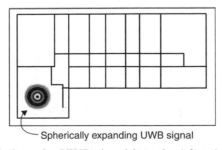

Figure 6.11   Simulation of a UWB signal launch. After A. Lauer, A. Bahr and I. Wolff, "FDTD simulations of indoor propagation", *44th IEEE VTC Conference Record*, Vol. **2**, June 1994, pp. 883–886.

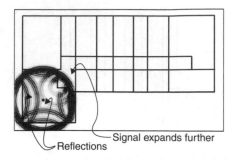

Signal expands further

Reflections

Figure 6.12 Simulated pulse transmission through walls. After A. Lauer, A. Bahr and I. Wolff, "FDTD simulations of indoor propagation", *44th IEEE VTC Conference Record*, Vol. **2**, June 1994, pp. 883–886.

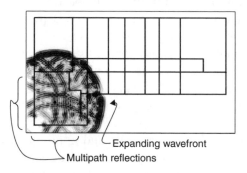

Expanding wavefront

Multipath reflections

Figure 6.13 Simulated multipath echoes. After A. Lauer, A. Bahr and I. Wolff, "FDTD simulations of indoor propagation", *44th IEEE VTC Conference Record*, Vol. **2**, June 1994, pp. 883–886.

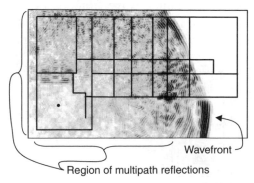

Wavefront

Region of multipath reflections

Figure 6.14 Simulated multipath echoes after wave-front passage. After A. Lauer, A. Bahr and I. Wolff, "FDTD simulations of indoor propagation", *44th IEEE VTC Conference Record*, Vol. **2**, June 1994, pp. 883–886.

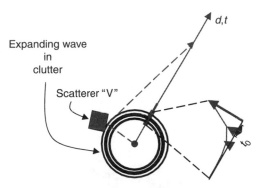

Figure 6.15 Simulated UWB impulse reaching an object.

the office area will have encountered an initial impulse as the wave front passes, followed by many overlapping echoes due to signals arriving from multiple reflected paths.

These wave-front propagation simulations were carried out using the Finite Difference Time Domain (FDTD) method (see [Kunz 1993]) by A. Lauer, and were also reported by [Zollinger 2002]. A UWB signal at any instant of time persists for only a short distance as seen by the dark bands of energy in Figures 6.11 through 6.14.

Figure 6.15 shows a UWB impulse wave front that has just reached an object that will reflect and scatter the signal. An amplitude profile of the UWB pulse is shown to the right. In Figure 6.16, the wave front, now

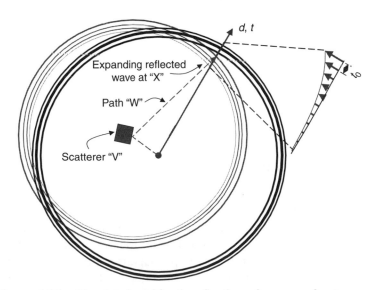

Figure 6.16 Simulated multipath reflection after wave-front passage.

at "X", has expanded, reflected, and scattered from the object "V" and has taken path "W." Both the original expanding wave and the reflected expanding wave interact at "X." The signal amplitude profiles of the signal prior to reflection from "V" in Figure 6.15, and at "X" in Figure 6.16, are shown on the side amplitude plots. A much more involved set of reflections follows the main expanding wave in Figures 6.11 through 6.14, but the principle is the same.

Delay Spread

6.3.2 Multipath and Delay Spread

Signals in multipath may be characterized by an echo-delay profile called the channel impulse response or CIR. The severity of the multipath is captured by the RMS delay-spread parameter τ_{RMS}, which is defined as the standard deviation (RMS or root-mean-square) value of the delay of reflections, weighted proportional to the energy in the reflected waves. Signals affected by multipath travel by paths of varying lengths from the transmitter to the receiver locale. If a transmitter sends an impulse $\delta(t)$, the received signal will be a collection of impulses

$$s(t) = \sum_n a_n \delta(t - \tau_n) \qquad (6.12c)$$

where the nth signal is delayed τ_n and has an amplitude a_n and an energy magnitude r_n. The impulse arrival time is characterized by a probability density function. The delay spread corresponds to its standard deviation, defined as

$$\tau_{RMS} = \sqrt{\frac{\sum r_n \tau_n^2}{\sum r_n} - \left[\frac{\sum r \tau_n}{\sum r_n}\right]^2} \qquad (6.12d)$$

A UWB signal propagating in a cluttered environment, such as inside a building, encounters many opportunities for reflections from surfaces and transmissions through walls and obstructions as seen in Figure 6.10. Energy-delay profiles resulting from propagation inside buildings tend to be exponentially distributed, and can be modeled by a sequence of impulses arriving at discrete times $t = nt_0$ at an average interval $t_0 = 1/B$

where B is the resolution bandwidth of the channel and is commensurate with the signal bandwidth f_B. The classic exponential energy profile [Jakes 1974] for an RMS delay spread τ_{RMS} comprises signal amplitude coefficients a_n defined in terms of r_n, which are uniformly distributed in $<-1, 1>$, thus,

$$a_n = \frac{r_n}{\sqrt{|r_n|}} \exp\left(\frac{-t_0 n}{2\tau_{RMS}}\right) \qquad (6.12e)$$

Defined in this way, $|r_n|$ is proportional to the signal energy. Signal amplitude a_n may have a positive or negative sign with the physical interpretation of having an even or odd number of polarity-inverting reflections. The multipath echo energy is not always distributed exponentially as in Equation (6.12e). Sometimes echoes arrive in clusters. One multipath channel model proposed by A. Saleh and R. Valenzuela (see [Saleh 1987]) accounts for such clustering and is one of the multipath models adopted along with the exponential model in the IEEE 802.15.3a channel model [IEEE 802 02/249] used for evaluating UWB systems. The IEEE 802.15.3a channel multipath model is summarized in Appendix B.

Figure 6.17 shows a calculated propagation-channel impulse response, that is, multipath signal coefficients for a 2-GHz bandwidth UWB signal. Multipath is modeled as impulses that are spaced uniformly at intervals t_0. The RMS delay spread in this example is $\tau_{RMS} = 5$ ns. The first and 20th coefficients and pulses are shown. The first pulse signal shown in Figure 6.17 is, of course, the only one that would be present in free-space propagation, because free-space propagation involves no reflections.

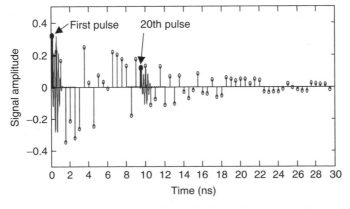

Figure 6.17 Calculated multipath coefficients for a signal in multipath.

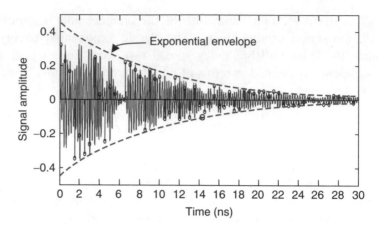

Figure 6.18   Calculated signal amplitude in multipath.

Figure 6.18 shows all of these particular impulse response coefficients multiplied by delayed copies of the radiated signal $g_0(t)$. The average of a large number of such randomly generated traces would exactly fill the exponential envelope.

6.3.3 UWB Signals Propagating in Multipath

A theoretical multislope wave propagation model derived by Siwiak, Bertoni, and Yano (SBY model) is described on the basis of ultra-wideband radio channel impulse response measurements in a dispersive channel. The theory, which applies to radio wave as well as to acoustic propagation, relates multipath delay spread, propagation law, and the maximum possible rake gain for the multipath propagation channel. Recent high-resolution UWB CIR measurements reveal different exponents in the distance dependence of the small area average for the total received energy, and the energy of the strongest arriving pulse. Additionally, the delay spread was found to increase with distance. A simple theory is described in this section to connect the differences in the propagation exponent with multipath scattering. This result has implications on receiving signals in multipath, rake gain, and understanding the multipath wireless channel.

The SBY propagation model [Siwiak 2003] described here is based on propagation measurements carried out in a large, single-story building using a UWB pulse transmitter and a UWB scanning receiver [Yano 2002]. The measured data were processed using the

CLEAN algorithm [Högbom 1974] to extract the amplitude of the
"strongest pulse" received in the energy-delay profile, as well as the total
energy in the profile and the RMS delay spread of the profile.

Scatter plots of the amplitude in decibels of the strongest pulse and
the total energy versus distance on a log scale between the transmitter
and the receiver are shown in Figures 6.19 and 6.20 respectively. The
scatter plot of RMS delay spread τ_{RMS} versus distance on a linear scale is
shown in Figure 6.21. The *least-squares-fit* line to the scatter plot, which
is also shown in Figure 6.19, indicates that the energy density of the
strongest pulse $P_S(d)$ has a dependence on the distance $d > 1$ between
the transmitter and receiver of the form

$$P_S(d) = \frac{P_S(1)}{d^3} \qquad (6.13)$$

having range index $n_S = 3$. By comparison, the least-squares-fit line to the
total energy in the delay profile, which is shown in Figure 6.20, indicates
a range index $n = 2$. Thus, the density of the total received energy for
effective radiated energy P_{EIRP} is given by

$$P(d) = \frac{P_{EIRP}}{4\pi d^2} \qquad (6.14)$$

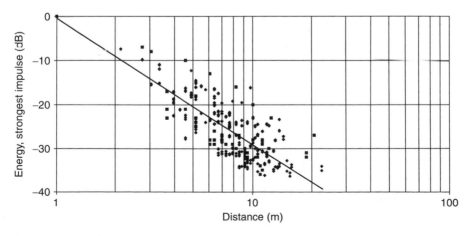

Figure 6.19 Strongest ray (impulse) energy density versus distance.
——— Median, curve fit; ◆ data points
Source: K. Siwiak, H. Bertoni and S. Yano, "On the relation between multipath
and wave propagation attenuation", *Electronic Letters*, **39**(1), 142–143, 2003.

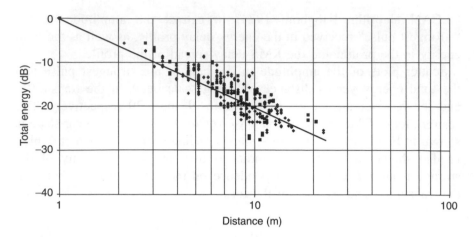

Figure 6.20 The total energy density versus distance exhibits $1/r^2$ behaviour.
—— Median, curve fit; ♦ data points
Source: K. Siwiak, H. Bertoni and S. Yano, "On the relation between multipath and wave propagation attenuation", *Electronic Letters*, **39**(1), 142–143, 2003.

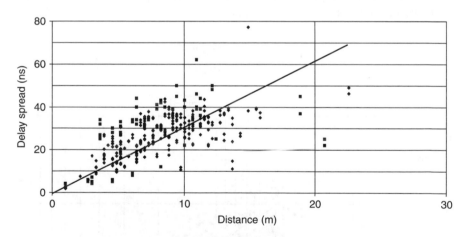

Figure 6.21 RMS delay spread versus distance.
—— Median, curve fit; ♦ data points
Source: K. Siwiak, H. Bertoni and S. Yano, "On the relation between multipath and wave propagation attenuation", *Electronic Letters*, **39**(1), 142–143, 2003.

Finally, the distance dependence of the RMS delay spread τ_{RMS}, which is given by the least-squares-fit line in Figure 6.21, is

$$\tau_{RMS}(d) = \tau_D d \tag{6.15}$$

where $\tau_D = 3$ ns/m. Ultra-wideband pulses propagating through a physical environment, such as a large building, are scattered by the many obstacles they encounter. Creation of these multipath pulses, as seen in Figure 6.22, removes energy from the primary pulse. The multipath pulses arrive at a receiver spread out in time, thereby creating the received time-delay profile (received energy vs. time). Integrating the received energy over the entire delay profile gives the total received energy. The time resolution of UWB technology allows the separation of the direct pulse, which is the first to arrive and is usually the strongest pulse, from the rest of the received pulses.

Because of the energy removed from the direct pulse by scattering, it is to be expected that the intensity of the direct pulse will decrease more rapidly with distance than the $1/d^2$ dependence in free space. At least for moderate distances d, the total energy flux will not be diminished by conversion to heat in the material encountered, and the energy is conserved. Thus, in a full 3D scattering environment such as a large building, conservation of the energy flux leads to the dependence $1/d^2$, seen in Figure 6.20 as 20 dB attenuation per distance decade.

Energy-delay profiles resulting from propagation inside buildings typically are a sequence of impulses arriving at discrete times $t = nt_0$ at an

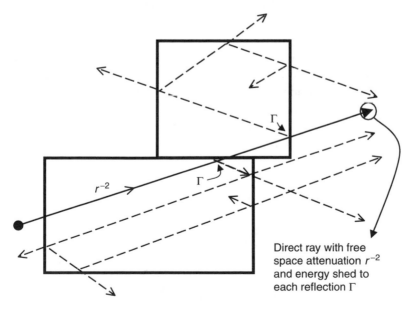

Direct ray with free
space attenuation r^{-2}
and energy shed to
each reflection Γ

Figure 6.22 The direct path as well as the reflected paths attenuate by geometrical expansion and shed energy to further reflections.

average interval $t_0 \ll \tau_{RMS}$. They often have, on an average, the classic exponential profile [Jakes 1974] of the form

$$S(d, t) = P_S(d) \sum_{n=0}^{\infty} \exp\left(\frac{-nt_0}{\tau_{RMS}}\right) \delta(t - nt_0) \qquad (6.16)$$

The total energy $P(d)$ is found by integrating $S(d, t)$ over all time, which leads to the relation

$$P_S(d) = P(d)\left[1 - \exp\left(\frac{-t_0}{\tau_{RMS}}\right)\right] \qquad (6.17)$$

between the total energy, the delay spread and the strongest arrival energy. The value t_0/τ_D represents a breakpoint distance d_t in the range index and equals 1 in Figure 6.19. For $t_0/\tau_{RMS} < 1$, or $d > 1$ in Figure 6.19, the term in brackets can be replaced by the negative of the argument of the exponential in (6.17), and

$$P_S(d) = \frac{P(d)t_0}{\tau_{RMS}} = \frac{P(d)d_t}{d} \qquad (6.18)$$

This relation is satisfied by the distance dependencies observed for the measurements made in a large building, where $P(d)$ varies as $1/d^2$, τ_{RMS} varies as d and $P_S(d)$ varies as $1/d^3$.

For propagation in urban environments from base-station antennas to mobiles, the average delay spread is thought to vary as $\tau_{RMS} = \tau_C d^{0.5}$, where d is in km and τ_C, which is the average delay spread at 1 km, has a value in the range 0.4–$1.0\,\mu s$ [Greenstein 1997]. For such radio links, the scattering environment is essentially a two-dimensional layer of finite thickness, so that the power scattered upward is lost. As a result, the average of the total power density in the scattering layer exhibits a distance dependence $1/d^n$ with $n > 2$. Equations (6.17) and (6.18) suggest that the strongest arriving pulse will have distance dependence $1/d^{(n+0.5)}$.

Rake Gain

6.3.4 Relation to Maximum Rake Gain

The maximum available rake gain is defined by the ratio of "total energy density" to "single-impulse energy density". When expressed in dB, and using Equation (6.17), on average

$$G_{\max} = 10 \log \left[\frac{P(d)}{P_S(d)} \right] = -10 \log \left[1 - \exp \left(\frac{-t_0}{\tau_{\mathrm{RMS}}} \right) \right] \qquad (6.19)$$

On the basis of indoor measurements, and for $d > 1$, $G_{\max} = 10 \log(d/d_t)$, so that $G_{\max}$ increases with d as $10 \log(d)$. However, in the case of outdoor systems, the dependence of τ_{RMS} of $d^{0.5}$ will result in $G_{\max}$ increasing only as $5 \log(d)$.

6.3.5 The SBY Median Multipath Propagation Model

A propagation model for the strongest impulse is based on the theory of Equation (6.17) with $t_0/\tau_{\mathrm{RMS}}(d) = (d_t/d)^{n-2}$, where the range index beyond d_t is n. Path gain P_G between 0-dBi antennas is weighted by receiver–antenna aperture $c^2/4\pi f_m^2$, where f_m is the geometric mean of the low- and high-frequency band edges of the UWB pulse, and c is the velocity of propagation. The path gain is described by

$$P_G = 10 \log \left\{ \left[\frac{c}{4\pi d f_m} \right]^2 \left[1 - \exp \left(- \left(\frac{d_t}{d} \right)^{n-2} \right) \right] \right\} \qquad (6.20)$$

Equation (6.20) is especially useful for modeling propagation in short-range indoor personal-area networks as exemplified by the IEEE 802.15.3 standard [IEEE802 2002]. It is also applicable to other scenarios that are typified by two distinct propagation laws.

Specifically, Equation (6.20) may be used as a simplified propagation model for the propagation between two antennas near a planar ground. Parameter d_t in Equation (6.20) is given by Equation (6.6), and the propagation law beyond that breakpoint distance is found from Equation (6.5) to be $n = 4$.

Figure 6.23 shows the two-path propagation for the specific signals discussed in Section 6.2, as well as square law and 4th power law asymptotes, which intersect at the breakpoint distance d_t. The multipath model of Equation (6.20) transitions smoothly between these asymptotes and can

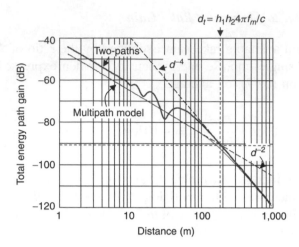

Figure 6.23   Comparison of propagation models.

be a reasonable choice of a propagation model when the specific signal shape is not available or cannot otherwise be determined.

6.3.6 Shadowing Variation and Statistical Link Design

A study of Figures 6.19 and 6.20 reveals that the signal strength at any given distance has a spread around the median value. At any particular point, the signal level has been attenuated by the geometrical spreading of an expanding spherical wave, but it has also undergone a large number of transmissions *through* and reflections *from* objects. The cumulative effect of these random multiplications by transmission and reflection coefficients is equivalent to the addition of random numbers in the logarithm of signal strength. The random events acting in multiplicative sequence produce the lognormal variation of slow fading. Multiplying random variables is equivalent to adding their logarithms and, by the Central Limit Theorem, so long as the chance of no one single event is dominant, the sum goes into a Normal distribution, and hence exhibits Gaussian statistics. This larger scale variation, or shadowing variation, is characterized by a lognormal standard deviation, which typically is in the range of $\sigma = 3$ to $6\,\mathrm{dB}$ for propagation within buildings or in an urban area. The probability P that a Gaussian random variable z is less than or equal to an arbitrary value Z, is found from

$$P\{z \leq Z\} = 1 - P_S = \frac{1}{2} - \frac{1}{2}\mathrm{erf}\left[\frac{z}{\sqrt{2}}\right] \qquad (6.21)$$

Table 6.3 Link-success probability versus z factor.

P_S	z
0.50	0
0.60	0.253
0.70	0.524
0.80	0.842
0.90	1.282
0.95	1.645
0.99	2.326

When applied to the large-scale signal distribution envelope, σ is the standard deviation in decibels of the signal distribution; it is the spread in the data points of Figures 6.19 and 6.20. The required signal margin is then

$$M = z\sigma \qquad (6.22)$$

P_S is the desired link-success probability, while $\text{erf}(u)$ is the error function. For a reasonable link reliability, we might desire $P_S = 0.90$. Table 6.3 lists values of z for several link-success probabilities P_S. Note that $z = 0$, that is, omitting the shadowing term is equivalent to accepting a 50% link-success probability. The basic propagation behavior is after all the *median* performance level. With a shadowing standard deviation of $\sigma = 4$ dB and a desired 95% link-success probability, at least $1.645\sigma = 6.6$ dB of additional link margin over the median level is needed.

Figure 6.24 shows the signal energy level exceeded by 95% of the lognormally distributed points compared with the median or 50% level. The median signal level and the 95% level are 6.6 dB apart when σ is 4 dB.

6.3.7 *Propagation Models and Parameters*

The propagation models are summarized in Table 6.4 along with their governing equations and parameters. Implicit in the models is the use of "constant gain" type of receiver antenna, that is, the receiver antenna has a collecting aperture that diminishes as the square of frequency, hence the apparent frequency dependency in the free-space model.

The parameters of the propagation models are summarized in Table 6.5 along with their commonly used units.

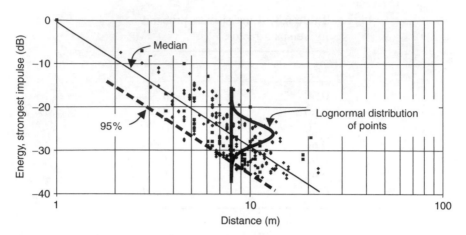

Figure 6.24   The median and 95th percentile signal levels.

Table 6.4   Propagation models and their applicability.

Model	Type	Governing equations	Parameters	Applicability
Free space	Deterministic	(6.1)	f_m	Free space
Two-path	Deterministic	(6.1), (6.12b)	f_m, H_1, H_2	Near ground, no clutter, signal defined
Two-path approximated by SBY	Statistical	(6.20)	f_m, d_t, $n = 4$	Near ground, no clutter, used when the signal shape is not known
SBY model with CIR and shadowing	Statistical	(6.20), (6.21), (6.22), (6.12c)	f_m, d_t, n, σ, τ_{RMS}	Full multipath with shadowing, and signal profile
IEEE 802.15.3a	CIR only		See Appendix B	

6.4 Summary

Propagation models useful for UWB signals were explained, including the free-space and two-ray deterministic models, and the SBY statistical multipath model. The free-space law is exactly the same for time-harmonic (sinusoidal, or narrowband) signals as it is for UWB signals. The two-path

Table 6.5   Propagation model parameters.

Parameter	Description	Units
f_m	Median frequency in propagating signal	GHz
H_1, H_2	Transmitting and receiving antenna heights above plane earth	m
d_t	Breakpoint distance between free-space law and n propagation law	m
n	Propagation power law	–
σ	Lognormal shadowing standard deviation	dB
τ_{RMS}	RMS delay spread	ns
t_0	Multipath resolution, reciprocal of signal bandwidth	ps

model signal shape is path length and frequency dependent. This law is applicable to propagation between antennas that are generally in the clear, and the signal paths can be accurately described by an unobstructed direct path plus a ground-reflected path. The SBY model is statistically based and provides a description of an average behavior of signals in full multipath. With the proper choice of parameters, the SBY model can be applied as a simplified model for two signal paths. The SBY model provides a theoretical basis for the propagation exponent in scattering and multipath. It reveals a close relationship between the propagation law, multipath delay spread, and rake gain. The model is particularly applicable to propagation in short-range wireless personal-area networks, and especially for UWB signals. A detailed short-range UWB channel impulse response (CIR) model was developed by the IEEEP 802.15.3a task group for evaluating the performance of UWB physical-layer proposals. The multipath component is based on a Saleh–Valenzuela model of exponentially distributed signal clusters. A statistical link design for reliability is possible, when, in addition to the multipath description, a shadowing standard deviation is also specified.

References

[Foerster 2001] J. Foerster, "The effects of multipath interference on UWB performance in an indoor wireless channel", *Proc. of the IEEE Vehicular Technology Conf*, Rhodes, Greece, May 2001.

[Friis 1946] H. T. Friis, "A note on a simple transmission formula", *Proceedings of the IRE*, **34**(5), 254–256; **41**, 1946.

[Greenstein 1997] L. J. Greenstein, V. Erceg, Y. S. Yeh and M. V. Clark, "A new path-gain/delay-spread propagation model for digital cellular channels", *IEEE Transactions on Vehicular Technology*, **46**(2), 477–486, 1997.

[Högbom 1974] J. A. Högbom, "Aperture synthesis with a non-regular distribution of interferometer baselines", *Astronomy and Astrophysics Supplement Series*, **15**, 417–426, 1974.

[IEEE802 02/249] *Channel Modeling Sub-committee Report – Final*, IEEE P802.15 Working Group for Wireless Personal Area Networks (WPANs), IEEE Document P802.15-02/249r0-SG3a, (Online): http://grouper.ieee.org/groups/802/15/pub/2002/Nov02/, December, 2002.

[IEEE802 02/301] *UWB Propagation Phenomena*, IEEE802.15 Document, (Online): <http://grouper.ieee.org/groups/802/15/pub/2002/Jul02/> document submission <02301r3P802-15_SG3a-UWB-Propagation-Phenomena.ppt>.

[IEEE802 2002] IEEE 802.15 Working Group for WPAN, (Online): <http://grouper.ieee.org/groups/802/15/>, 31 July 2002.

[Jakes 1974] W. C. Jakes, *Microwave Mobile Communications*, IEEE Press, Piscataway, NJ: 1974, Reprint.

[Kunz 1993] K. S. Kunz and R. J. Leubbers, *The Finite Difference Time Domain Method for Electromagnetics*, Boca Raton, FL: CRC Press, 1993.

[Lauer 1994] A. Lauer, A. Bahr and I. Wolff, "FDTD simulations of indoor propagation", *44th IEEE VTC Conference Record*, Vol. **2**, June 1994, pp. 883–886.

[McKeown 2003] D. McKeown, *Gammz UWB Cartoons and Art*, Private Communication to K. Siwiak, December 2003.

[Saleh 1987] A. Saleh and R. Valenzuela, "A statistical model for indoor multipath propagation", *IEEE JSAC*, **SAC-5**(2), 128–137, 1987.

[Siwiak 1998] K. Siwiak, *Radiowave Propagation and Antennas for Personal Communications*, Second Edition, Norwood, MA: Artech House, 1998.

[Siwiak 2001] K. Siwiak, "The future of UWB – a fusion of high capacity wireless with precision tracking", *Forum on Ultra-Wide Band*, Hillsboro, OR, (Online): <http://www.ieee.or.com/IEEE ProgramCommittee/uwb/uwb.html>, 11–12 October 2001.

[Siwiak 2001a] K. Siwiak and L. L. Huckabee, "An introduction to ultra-wide band wireless technology", in B. Bing (Ed.), *Wireless Local Area Networks – The New Wireless Revolution*, New York: Wiley, 2001.

[Siwiak 2003] K. Siwiak, H. Bertoni and S. Yano, "On the relation between multipath and wave propagation attenuation", *Electronic Letters*, **39**(1), 142–143, 2003.

[Yano 2002] S. M. Yano, "Investigating the ultra-wideband indoor wireless channel", *Proc. IEEE VTC2002 Spring Conf.*, Vol. 3, Birmingham, AL, May 7–9, 2002, pp. 1200–1204.

[Zollinger 2002] E. Zollinger, *Radio Channel in UMTS-Systems*, IMST GmbH, (Online):   <http://www.imst.com/mobile/itg/itg_umts/zollinger.pdf> 8 April 2002.

7

Receiving UWB Signals

Introduction

Signals, once generated, transmitted, and propagated, must be received in order to be understood. A receiver, such as your home entertainment radio, takes in the signals and translates them to music. Receivers carry out several functions. They capture signal energy with the antenna, extract the information from the signal, and present that information. The art is in taking the signal in and *efficiently* recovering the conveyed information. We have already been introduced to early radio with its wideband noisy signals and primitive receiving and inefficient detecting devices. Efficient signal recovery and information detection is the key to a successful radio link. Efficient signaling involves *matching* the signal to the receiver and detector.

Marconi's transmission of the letter "S" across the Atlantic did not convey a great deal of information. The transmission was prearranged. The prearranged signal enabled the use of a simple "matched receiver" at the reception point. More accurately, it was a matched receiving technique: the receiving operator knew to listen for the prearranged, predetermined letter "S" in Morse code. He also knew when to expect the signal. This greatly enhanced the possibility of receiving a weak signal buried in background static and noise. Knowing *when* to receive something and *what* the signal actually comprises makes it an expected event, and makes its detection a lot simpler than stumbling on an intelligent but *random* event. It is a technique that Morse telegraphists regularly apply when using the rich set of internationally agreed-upon abbreviations and shortcuts available in Morse telegraphy. Each radio service today, such as public safety, aviation, and so on, have developed standard phrases and

Ultra-Wideband Radio Technology   Kazimierz Siwiak and Debra McKeown
© 2004 John Wiley & Sons, Ltd   ISBN: 0-470-85931-8

expected sequences of specific information presented in an expected order to enhance correct reception. Radio develops its characteristic jargon for the purpose of enhancing clarity of transmission. The "characteristic jargon" in UWB signal reception is the "template" signal that we attempt to match to the received energy. We will see that the efficiency with which we can recover our desired signal depends on the quality of that match, both in timing and in signal shape. The few and simple concepts regarding the efficient reception and detection of UWB signals should not obscure the reality that efficient receiver design is a major engineering challenge. Our goal here is the exposition of the relatively simple concepts rather than their challenging implementation.

7.1 Reception of UWB Signals

Receiving a UWB signal is a matter of "matching" the received energy to a predetermined template. When we know the expected shape of the signal and we know something about when the signal is expected, shape and time are matched; we enhance our ability to receive the signal and its information content. Time gating, the turning on of the receiver and detector for just the expected duration of the expected signal is one technique of "matching," and we will look at the efficiency of this technique. If the gate is open too long, then excess noise energy will be collected, obscuring the signal energy and affecting the signal-to-noise ratio.

Time Gating

From the mathematics point of view, the template signal is multiplied by the received signal and the resulting product or "correlation" is collected as signal energy in an "integrator" or energy accumulator. Any mismatch between the signal shape and the template shape, or any misalignment in time between the two, results in a reduced efficiency in the collection of the energy. The signal is not as clear as it could possibly be because of the presence of noise.

7.2 Noise and Interference

Thermal noise, sometimes idealized as additive white Gaussian noise (AWGN), and "man-made" interference from other systems occupying the spectrum, place a limit on the range and on the capacity of a wireless system. Noise includes thermal noise caused by random vibrations of charges on a lossy conductor, interference, and meteorological noise. In personal wireless-network applications, thermal noise and interference play the most important roles. Natural and man-made noise sources vary with frequency in the radio spectrum as seen in Figure 7.1.

From quantum mechanics considerations, the noise-power density in watts per hertz generated in any lossy element is (see [Siwiak 1998])

$$N_0 = hf \left[\frac{1}{e^{hf/k_b T} - 1} + 1 \right] \tag{7.1}$$

where
$h = 6.6260693 \times 10^{-34}$ J · s is Plank's constant,
$k_b = 1.3806505 \times 10^{-23}$ J/K is Boltzmann's constant [CODATA 2002],
f is frequency in hertz (Hz) and
T is the absolute temperature in kelvin (K).

When frequency f is small enough, which includes all the cases that we are dealing with here, noise power density equation reduces simply to

$$N_0 = k_b T \tag{7.2}$$

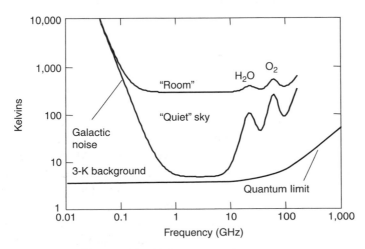

Figure 7.1 Noise versus frequency.

Noise *power* N is noise density multiplied by the bandwidth B under consideration so

$$N = k_b TB \qquad (7.3)$$

The wider the bandwidth, the more the noise power collected. When we were discussing the Hertzian and Marconi experiments, we noted that unnecessarily wide *signal* bandwidths were emitted, and to capture the emitted energy, wide receiver bandwidths were needed. The noise that was captured in addition to the signal was of course proportional to the signal bandwidth, which far exceeded the *information bandwidth*, hence the problem! The bottom line is that the noise power received is proportional to the bandwidth of the receiver, and it is therefore highly desirable to limit that bandwidth to the information bandwidth – something that was accomplished in early wireless systems by sharp, narrowband tuning circuits.

Band-Pass Filter

7.3 Receiver-detector Efficiency

In Chapter 4, we introduced a conceptual UWB transmitter and receiver system (see Figure 4.4), which showed early roots in spark-gap technology. We revisit that system now to investigate why it might perform badly. Early spark transmitters generated very wideband signals, which were much wider than the *information bandwidth* that they conveyed. For example, the Morse telegraphy employed with these systems at 25 words per minute conveys approximately 20 bps of information. Our "spark UWB" system, reproduced in Figure 7.2, was crafted to occupy a 500-MHz bandwidth. The simple amplitude detector receiver is similarly 500 MHz wide, but the information conveyed by the Morse coded signal is only 20 Hz wide. Thus, the mismatch between the transmitted signal and the way that the information is received and detected is 20/500,000,000. In decibels, this is an efficiency of *minus* 74 dB!

To be heard reliably, the signal must be another 6 dB or so stronger than the total noise received in that 500-MHz bandwidth. Early wireless went narrowband to match the signal bandwidth to the information bandwidth.

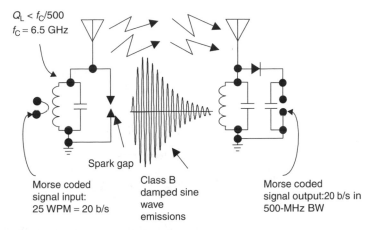

$Q_L < f_C/500$
$f_C = 6.5$ GHz

Spark gap

Morse coded
signal input:
25 WPM = 20 b/s

Class B
damped sine
wave
emissions

Morse coded
signal output:20 b/s in
500-MHz BW

Figure 7.2 An inefficient UWB signaling system based on early wireless technology.

Transmitters became "cleaner" and emitted "pure tone" signals, while receivers were tuned more narrowly to exclude as much noise as possible but still pass the information bandwidth. The signal-to-noise ratio S/N or SNR was greatly improved. The receiver-detector efficiency can be expressed as the output SNR divided by the input SNR. In digital systems rather than SNR, we use the data-bit energy-to-noise density ratio, E_b/N_0. SNR is most commonly an analog signal measure and is a ratio of *powers*, while E_b/N_0 is clearly associated with binary signaling and is a ratio of *energies*. Signal-to-noise ratio and bit energy-to-noise density ratio are related in AWGN by taking into account the bit duration T_b, and the signal bandwidth B,

$$SNR = \frac{S}{N} = \frac{(E_b/T_b)}{(N_0 B)} \qquad (7.4)$$

SNR and E_b/N_0 are equal when the signal bandwidth equals the inverse of the bit duration. In analog systems, B is often taken to mean the net bandwidth of all the receiver filters, and is often wider than the signal bandwidth. A simple receiver architecture for which the efficiency may be calculated is shown in Figure 7.3.

A signal collected by the antenna is first filtered. The filter has an impulse response equal to $h(t)$. This filtered signal is mixed with a template signal $p(t)$. For digital signals, we define the detector efficiency e_c by

$$e_c = 10 \, \log\left[\frac{(E_b/N_0)_{out}}{(E_b/N_0)_{in}}\right] \qquad (7.5)$$

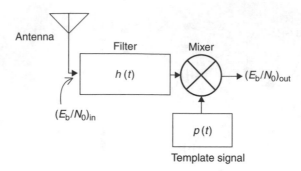

Figure 7.3   A receiver system.

An equivalent expression involving *SNR* can be written for an analog signal by using Equations (7.4) with (7.5). The efficiency can be maximized when the signal $s(t)$ supplied to the mixer is matched by the template signal $p(t)$ generated in the receiver. In this case, the filter is an "all pass" filter, that is, its impulse response is an impulse

$$h(t) = \delta(t) \text{ with } p(t) = s(t) \tag{7.6}$$

$\delta(t)$ is the Dirac delta function. This is known as a "matched template" receiver. Alternatively, a "matched filter" might be used. The filter impulse response in this case must equal a reverse-time copy of the received signal

$$h(t) = s(-t) \text{ with } p(t) = \delta(t) \tag{7.7}$$

Equations (7.6) and (7.7) give identical mathematical results: a perfect efficiency detector. There is an additional constraint on the filter in Equation (7.7). The filter must be *causal*. That is, it cannot have an output before it has an input. Matched filters might not always be physically realizable. There is a range of gradation between the matched filter and the matched template. In general, the receiver-detector implementation efficiency e_c (see [Siwiak 2001]) of this operation is

$$e_c = 10 \, \log \left[\frac{\left| \int \int s(\tau)h(\tau - t) \, d\tau p(t) \, dt \right|^2}{\int s(t)^2 \, dt \int \left| \int p(\tau)h(\tau - t) \, d\tau \right|^2 \, dt} \right] \tag{7.8}$$

and is maximized when

$$C \int p(\tau)h(\tau - t)\,\mathrm{d}\tau = s(t) \qquad (7.9)$$

provided $h(t)$ is causal. The constant C is the RMS value of $s(t)$. The correlator efficiency depends strongly on the shape of the signal $s(t)$ and its relationship to $h(t)$ and $p(t)$. The efficiency e_c can typically range from $-6\,\mathrm{dB}$ to about $-1\,\mathrm{dB}$ for simple templates, and might include rake gain of several dB for a self-correlating receiver. Equation (7.9) states that the convolution of the filter response and the template shape must equal the signal shape for perfect efficiency. It also shows that the matched template and matched filter designs are not an either/or proposition, but that the filter and the template can be designed to cooperate in the quest for improved detector efficiency. One possible way of improving the efficiency of the wireless transmission system in Figure 7.2 is to use some form of signal matching at the receiver, as shown in Figure 7.4.

This system employs a correlating mixer in the receiver, followed by an integrator that acts like a data bandwidth filter. The template signal implied in Figure 7.4 is a sine wave centered at the signal center frequency. We recognize that particular receiver as a "homodyne" receiver, or in more modern terminology, a "zero-IF" receiver. There is a mismatch between the "damped sine wave" signal shape and a continuous sine wave template, but it is on the order of several decibels and is a considerable improvement over the 74-dB loss that we described earlier in relation to in the circuit of Figure 7.2.

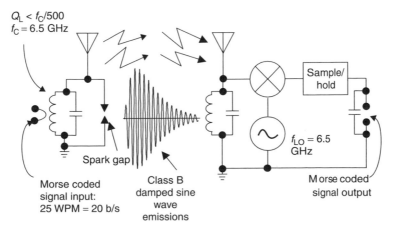

Figure 7.4 An efficient UWB signaling system using a matched receiver.

7.4 Efficiency of Simple Templates

In some UWB systems, for example, those employing PPM and time-modulation, it is convenient to use a simple, rectangular pulse for the correlator template signal. The detector efficiency can be determined by applying Equation (7.8). Figure 7.5 shows a 100% bandwidth UWB impulse signal $s(t)$ and a perfectly positioned, optimal-width rectangular template pulse $p(t)$. We first encountered this pulse shape earlier in Figure 5.11 in our study of impulse radiation. Now we will investigate the efficiency of its detection using a rectangular template.

The numerator in Equation (7.8) is physically implemented by a mixer circuit as shown in Figure 7.3. The denominator comprises signal normalizations. The efficiency of the template-signal match is shown in Figure 7.6 as a function of the template width. In Figure 7.7, the efficiency is shown for an optimal-width template and for templates that are 25% wider and narrower than the optimum width, but as a function of template positioning timing jitter error. A 25% error in template width costs less than 0.3 dB in receiver efficiency. Positioning the template in the wrong place causes additional loss. The curves show the average effect of a uniformly distributed positioning error, *timing jitter*, of the template relative to the optimum placement. As much as another 0.5 dB loss can result from the *timing errors*. This kind of error is analogous to a phase error in the local oscillator of a conventional "zero-IF" receiver design.

The net efficiency of a rectangular template and the wideband signal $s(t)$ shown in Figure 7.5 is 3.5 to 4 dB with reasonable assumptions on

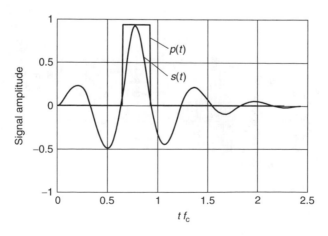

Figure 7.5   A UWB signal and a rectangular template.

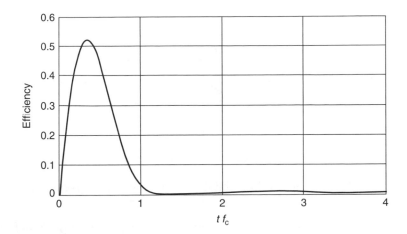

Figure 7.6   Efficiency as a function of template width.

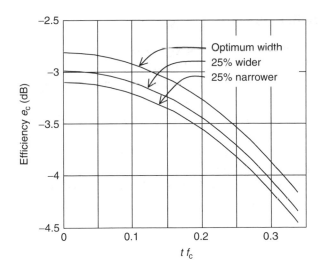

Figure 7.7   Efficiency with template-timing jitter and width errors.

template width and timing errors. The receiver-detector efficiency using rectangular templates decreases as the UWB pulses bandwidth decreases because of signal mismatch.

A simple conceptual receiver block diagram, as in Figure 7.3, should not belie the fact that considerable additional circuitry and signal processing is needed to lock to a UWB signal. Signals are coded, modulated, and they have bandwidths and center frequencies within a band. All of these

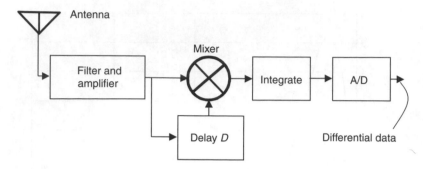

Figure 7.8   A differential (self-correlating) receiver.

need to be aligned for the efficient reception and detection of signals. All of these add cost and complexity to a UWB receiver just as they do to the analogous conventional receivers.

7.5 The Self-correlating Receiver

Perhaps the simplest UWB receiver is one operating on the transmitted reference (TR) principle. The TR receiver shown in Figure 7.8 comprises a filtered amplification stage following the antenna, and a mixer for multiplying the signal with a delayed copy of the signal. The product of the multiplication is integrated and sampled as output data. In its simplest form, a communications system requires that only pulse pairs be sent [Hoctor 2002]. Two pulses of opposite polarity may indicate a data "1" while pulses of the same polarity indicate a data "0." The integration interval is somewhat longer than a pulse length to rake in some of the multipath. Using pulse pairs is not energy efficient; however, it is easy to implement in a transmitter and in the receiver.

The self-correlating receiver may be used to detect the UWB equivalent of a conventional differential phase shift keying (DPSK). Data bits are sent differentially encoded and spaced at time intervals D. Differential encoding requires that an initial starting impulse be sent. Then a data "1" is meant if, in subsequent impulses, the pulse polarity changed from the previous pulse, and a "0" if the same polarity is sent. The integration spans the length of an impulse plus the delay time D.

7.6 Summary

We have explored the art of capturing the signal and *efficiently* recovering the energy that conveys the information. Relatively few simple concepts govern the efficient reception and detection of UWB signals. These should not obscure major engineering challenges that comprise the design of efficient and cost-effective receivers. Receiving a UWB signal is a matter of "matching" the received energy to a template. This template is signal-generated in the receiver, or can be transmitted along with the information signal. Efficient reception and detection involves matching the signal in shape as well as in time. Receiver and detector efficiency are important because received signal energy competes with received noise energy, and it is the SNR that determines successful information transfer.

References

[CODATA 2002] *2002 CODATA Recommended Values of the Fundamental Physics Constants*, 31 December 2002, (Online): <http://physics.nist.gov/cuu/Constants/>, December 2003.

[Hoctor 2002] R. T. Hoctor and H. W. Tomlinson, *An Overview of Delay-Hopped, Transmitted-Reference RF Communications*, General Electric Company, 2001CRD198, January 2002, (Online): <http://www.crd.ge.com/cooltechnologies/pdf/2001crd198.pdf>, 10 December 2003.

[McKeown 2003] D. McKeown, *Gammz UWB Cartoons and Art*, Private Communication to K. Siwiak, December 2003.

[Siwiak 1998] K. Siwiak, *Radiowave Propagation and Antennas for Personal Communications*, Second Edition, Norwood, MA: Artech House, 1998.

[Siwiak 2001] K. Siwiak, T. M. Babij and Z. Yang, "FDTD simulations of ultra-wideband impulse transmissions", *Proc. of IEEE Radio and Wireless Conf. – RAWCON 2001*, Boston, MA, August 19–22 2001.

8

UWB System Limits and Capacity

Introduction

UWB does not operate in a world of its own where it would be free to propagate without intrusion by the rest of the electromagnetic noise. Without other electromagnetic devices, regulations, and standards, UWB could, more ideally, operate almost perfectly. Even in this perfect world, UWB has limits – limits placed on it by the restrictions of our physical world.

In this chapter, we will see how noise, both thermal and man-made interference, limits wireless system performance. We will introduce Shannon's important theoretical formula, which expresses a restrictive tie between system capacity, bandwidth, and signal-to-noise ratio (SNR). Antenna characteristics, especially as energy-collecting apertures, also play a role in performance by bringing the operating frequency into consideration. We will find a simple way to express the performance of any wireless link in terms of fundamental physical limits, and without resorting to specific radio implementations. There is a theoretical maximum limit on UWB system gain, given the theoretical limits of Shannon as well as the power density and frequency limits imposed by regulations. These limits result in a "system gain" per bit per second of 173 dB/bps for operation within the US FCC regulations. The limiting considerations provide us with a simple means by which to estimate the performance and capabilities of practical radio links, particularly UWB links, without specifically resorting to the technological details.

Ultra-Wideband Radio Technology   Kazimierz Siwiak and Debra McKeown
© 2004 John Wiley & Sons, Ltd   ISBN: 0-470-85931-8

8.1 Limits in Communications

Communications links have certain resources available to them, and definite limits regarding access to these resources. Among the resources are transmitter power, signaling method, and receiving system. In fact, we will introduce a fundamental limit to the system link gain in terms of ideal access to the resources. We have, however, limited access to the resources. For example, noise impairs our received signals, the available bandwidth and power are under regulatory constraints, practical modulations have various efficiencies, and antenna characteristics impose constraints. Each of these limits, physical and regulatory, constrains the radio-link performance in a box. We will present the Fundamental System Gain Limit from which allocations are made for system parameters and system inefficiencies. The remainder, we will see, is allocated for propagation losses.

8.1.1 Noise

We developed the concept of thermal noise, AWGN, in the previous chapter. Noise and interference from other systems occupying the spectrum place a limit on our ability to extract information from signals. In fact, noise places limits on the *range* and on the *capacity* of a wireless system. More precisely, the ratio of the desired signal power to noise power ultimately limits performance. In certain systems, performance can be limited by thermal noise, while others are often limited by interference, either from the same system of radios or from other systems. The distinction can be important, as the receiver sensitivity is often a costly parameter critical in thermal noise-limited systems (such as in satellite links), but somewhat less important in interference-limited systems. We will revisit this when we look at system capacity. For now, one of our UWB system limitations is noise power, N_0B, introduced in Chapter 7. The noise power, we recall, is

$$N_0B = k_bTB \tag{8.1}$$

where $k_b = 1.3806505 \times 10^{-23}$ J/K is Boltzmann's constant [CODATA 2002], T is the absolute temperature in kelvin (K), and B is the bandwidth in hertz (Hz). The noise power in a 1-Hz bandwidth, using Equation (8.1), is $N_1 = -204$ dBW or -174 dBm at room temperature. This is the first of our physical limits.

8.1.2 Shannon's Capacity Formula

It will be instructive to know the theoretical limits of any system so that we can better understand the efficiency of our UWB communication system. Already we have learned that there is an advantage in having wide bandwidths available and wideband signals to convey information. First, we will review how bandwidth impacts a UWB communication link.

Let us introduce a term e_b to denote the *relative efficiency*, or *communications efficiency*, of a particular communications system and define it as the minimum signal energy per data bit E_b to noise power density ratio N_0 [Lindsey 1973]

$$e_b = 10 \ \log \left(\frac{E_b}{N_0} \right) \tag{8.2}$$

required to achieve a sufficiently small data error rate. Thus, the smaller the e_b, the better our system will perform in AWGN. We want to find a limiting value for the relative efficiency, the smallest value of e_b that is physically possible. This value can be derived from the theoretical work of Claude Shannon [Shannon 1948]. Shannon derived a relationship between the channel capacity C, the channel bandwidth B, and the signal-to-noise ratio S/N:

$$C = B \ \log_2 \left(1 + \frac{S}{N} \right) \tag{8.3}$$

In the simplest terms, the information capacity of a communications system is proportional to the bandwidth B that is employed. Once the bandwidth is fully utilized, any further increase in capacity comes at an exponential increase in the required signal-to-noise ratio. Just invert the formula solving for the required signal-to-noise ratio

$$\frac{S}{N} = 2^{C/B} - 1 \tag{8.4}$$

in terms of the available bandwidth and the capacity of the system. This relationship is shown in Figure 8.1.

Bandwidth-constrained systems can achieve large capacities. Plain old telephone system (POTS) voice lines guarantee a bandwidth of less than 3 kHz, yet we use 56-kbps data modems quite successfully over the lines. Not counting modem compression techniques, we obtain the higher data rate performance when the SNR over the line is sufficiently high, greater

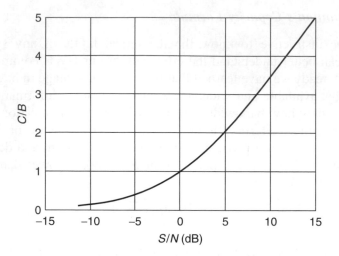

Figure 8.1   The ideal channel performance.

than $2^{56k/3k} - 1 = 416,000{:}1$ (56 dB). A line that has a bandwidth of 56 kHz would, by comparison, require an SNR of just 1 (0 dB). Thus, with a given noise-power level, the bandwidth-constrained line needs 416,000 times as much signal power as the 56-kHz line for the same level of performance. In the case of UWB, 7.5 GHz of bandwidth is available (in the United States), so exceptionally high data rates are possible at signal-to-noise ratios depending only on the modulation efficiency. UWB can operate in the linear Shannon region for C up to $B = 7.5$ GHz.

Let us repeat this in the language of Shannon. Signals convey information to the extent they provide *unexpected* data. Expected results do not convey a great deal of information. As previously discussed, Marconi's prearranged transmission of the letter "S" across the Atlantic Ocean conveyed little more "information" than the verification that his wireless link worked over the distance. The information content of a sufficiently coded signal is *entropy* – it is a measure of how well the energy is spread out. Another way of looking at a stream of unexpected bits is in comparison to random data, which resembles random noise. Shannon showed that the more a transmission resembles random noise, the more information can be conveyed, provided it is modulated to a regular carrier frequency such as a sine wave signal. In UWB, the regular carrier is the UWB impulse. One needs a low-entropy carrier to bear a high-entropy message. Hence, according to Shannon, an alternative to transmitting a signal with a high power density and low bandwidth would be to use a low power density and a wide bandwidth. We are approaching the notion that signal spreading has the operational advantage of being resistant to interference. This plays to an advantage of UWB – large available bandwidth.

Shannon states that there exists an optimum coding of signals, not restricted in bandwidth, that makes it possible to communicate data error free over an AWGN channel, provided that the system data rate R is less than or equal to the channel capacity C. Now, we are ready to find the highest modulation efficiency (smallest value of e_b) that is physically possible for error-free communication. We start by noting that the signal-to-noise ratio S/N and E_b/N_0 are related by

$$\frac{S}{N} = \frac{(E_b/N_0)}{(T_b B)} \tag{8.5}$$

where T_b is the time required to send one data bit. It is straightforward to show [Couch 1993] that when bandwidth B is unbounded (infinite), the capacity limits to

$$C = \frac{E_b}{(N_0 T_b B \ln(2))}. \tag{8.6}$$

Then, by applying Shannon's criterion, error-free communication is possible up to the limit that the information rate $1/T_b$ equals the channel capacity C. We get maximum possible communication efficiency

$$e_{b\,min} = 10 \, \log\left(\frac{E_b}{N_0}\right) = 10 \, \log(\ln(2)). \tag{8.7}$$

Expressed in decibels, $e_{b\,min} = 10\log(\ln(2)) = -1.59\,\text{dB}$ is the best communications efficiency possible in a bandwidth-unconstrained AWGN channel. It is a limit of nature and is our second physical limit. This is one of our key limiting values that defines the most efficient modulation possible. While theoretically laudable, Shannon leaves no recipe for practically achieving this efficiency. For our purposes, we note that different modulation schemes will achieve different communication efficiencies, and that some modulations and coding techniques, for example turbo codes, can approach the theoretical limit.

8.1.3 Communication Efficiency of Various Modulations

In Chapter 4, we showed that UWB signals may be modulated, that is, have information encoded on them in many ways. We identified a range of UWB modulations exemplified by

1. Pulse Position Modulation (PPM)

2. *M-ary* Bi-Orthogonal Keying (*M*-BOK) Modulation

3. Pulse Amplitude Modulation (PAM)

4. Transmitted Reference (TR) Modulation.

Let us look at the relative efficiencies of these modulations. We can compare the relative modulation efficiencies at several different bit error rates (BER) in Table 8.1. The modulations are listed roughly in decreasing efficiency. Two-level BOK and PAM (equivalent to conventional BPSK) and a BER $= 10^{-3}$ are highlighted as a convenient reference for modulation efficiency.

Modulation complexity can be increased to combine both pulse amplitude and pulse position to form the UWB equivalent of conventional *M*-ary QAM (Quadrature Amplitude Modulation). The *M*-BOK family of modulations has the interesting property that, as the modulation complexity, *M*, increases, the modulation efficiency tends toward the

Table 8.1 Modulation efficiencies.

Modulation	Modulation efficiency, e_b (dB)			
	BER $= 10^{-2}$	BER $= 10^{-3}$	BER $= 10^{-5}$	
64-BOK	2.4	4.1	6.1	
16-BOK	3.0	4.9	7.1	
8-BOK	3.4	5.4	7.8	Equations (4.14), (4.15)
4-BOK	3.8	6.1	8.6	
2-BOK/ 2-PAM/ BPSK	4.3	6.8	9.6	Equation (4.16)
PPM/OOK	7.3	9.8	12.6	Equation (4.13)
N-TR	5.9	7.9	10.3	Equation (4.19)
2-TR	8.9	10.9	13.3	
4-PAM	8.3	10.8	13.5	
8-PAM	12.8	15.2	18.0	Equations (4.17), (4.18)
16-PAM	17.6	20.1	22.9	

Table 8.2 Modulation efficiencies relative to ideal.

Modulation	Modulation efficiency relative to $e_{b\,min}$ at BER $= 10^{-3}$
64-BOK	5.7
16-BOK	6.5
8-BOK	7.0
4-BOK	7.7
2-BOK/2-PAM/BPSK	8.4
PPM/OOK	11.4
N-TR	9.5
2-TR	12.5
4-PAM	12.4
8-PAM	16.8
16-PAM	21.7

Shannon-limited value of $e_{b\,min} = -1.59\,dB$. On the other hand, the efficiency performance of amplitude-encoded modulations such as M-PAM and M-QAM will appear below our BPSK reference modulation.

Table 8.2 presents the modulation efficiency relative to the ideal efficiency for the various modulations operating at a 10^{-3} BER.

Amplitude encoding of data tends to result in inefficient modulations because it operates on the S/N term in Shannon's Equation (8.3). Orthogonal and bi-orthogonal encoding and modulation schemes, conversely, tend to make the signal more noiselike and hence can more efficiently utilize the linear relationship in Equation (8.3) between capacity C and bandwidth B up to that available system bandwidth. The modulations of Table 8.1 are by no means exhaustive. Many variations are possible, including methods of encoding pulses in frequency, time-position, polarity, and amplitude (see, for example, the frequency/time-position modulation in [IEEE802 03/105]) as well as coding methods like turbo codes. Modulation efficiency is one of the fundamental limitations in UWB radio performance.

8.1.4 Regulatory Limits

A communications link requires signal power. In UWB, there are two bounds that limit the available power. One is the power spectral density limit measured in dBm/MHz, and the other is the frequency-band limit.

Table 8.3 UWB EIRP limits in various jurisdictions.

Jurisdiction	UWB range (GHz)	Emission level (dBm/MHz)	Maximum EIRP, (dBm)
United States	3.1–10.6	−41.3	−2.55
Singapore UFZ	2.2–10.6	−35.3	+3.94
ETSI indoor [a]	3.1–10.6	−41.3	−2.55
ETSI handheld [a]	3.1–10.6	−61.3	−22.55

[a] As of this printing, the ETSI limits are proposals.

Together they define the maximum possible radiated power for a UWB signal. Practical UWB implementations will use some fraction of this maximum. This is the third limiting parameter for UWB systems, and its value is determined by the various regulatory jurisdictions. Table 8.3 lists some of the regulatory limits, and the maximum possible EIRP if all the bandwidth is utilized.

The Singapore values are within the UWB Friendly Zone (UFZ), and the ETSI values are under consideration at the time of this printing.

8.1.5 Antenna Apertures and Propagation

Antennas can play a subtle role in defining the effectiveness of a UWB link. The large UWB bandwidth means that energy in the lower or upper frequency limits might be captured differently by a receiving antenna depending on the antenna type. Antennas are characterized by a receiving aperture area, which is directly related to antenna gain. Over the frequency band, an antenna might exhibit "constant-aperture" or "constant-gain" characteristics, or it might be in between these two limits. The distinction is important, because antenna choice can render one end of the band more effective than the other end of the band in a UWB link.

The aperture of a *constant-gain* antenna remains constant in units of wavelength. For instance, a dipole antenna has an aperture of approximately 0.13 square wavelengths and a power gain of about 1.5 at any frequency. As frequency increases, the wavelength decreases, and the constant gain–antenna aperture decreases as the inverse square of frequency. This is typical of omnidirectional antennas, which are designed to have a constant gain and pattern. Thus, omnidirectional antennas exhibit this behavior. A *constant-aperture* antenna is one whose antenna-aperture area remains fixed with frequency. For instance, a horn antenna will typically (but not always) have a fixed aperture. As frequency increases, the size of this aperture in units of wavelengths increases as the square of frequency. This increases the antenna gain also as the square of frequency. Many (but not all) directive antennas exhibit this behavior. The penalty with constant-aperture antennas is that the beam width decreases as frequency increases, thus narrowing the antenna's field of view. In a UWB link having a constant-gain receiving antenna, the received signal power decreases as the inverse square of frequency. A link using a constant-aperture receive antenna captures signal energy equally well anywhere in the UWB band, but at the price of a narrowing field of view as the frequency increases. Either type of antenna may be used on the transmitter side with equal effect because the transmitter signal may be compensated to ensure that the emitted power meets the same flat EIRP spectral mask permitted under the regulations [Schantz 2003].

The *effective aperture* A_e of an antenna is related to the gain G by

$$A_e = \frac{\lambda^2}{4\pi} G \qquad (8.8)$$

where $\lambda = c/f$ is the operating wavelength. The effective aperture and the actual size of the antenna are not necessarily the same: a half-wave long dipole, for example, might typically have a diameter of less (often much less in narrowband applications) than a tenth of a wavelength for a total projected area of $0.05\lambda^2$ and exhibits a gain of $G = 1.6$ (2.1 dBi). Its effective aperture, however, is $(1.64/4\pi) \lambda^2 = 0.13\lambda^2$. The "excess" aperture is due to the near-field structure of the antenna. A point to note: any extraneous materials within the near-field structure of the antenna will tend to decrease the antenna efficiency, often by 3 to as much as 9 dB.

Consequently, the UWB system has an available parameter for trade-off. When directional antennas of the constant-aperture type can be used, the entire UWB spectrum has equal value from the propagation point of view. When omnidirectional properties are desired, then the collecting aperture area is proportional to the wavelength squared. Less signal is captured as the frequency is increased. For our analysis, we use an antenna aperture consistent with an omnidirectional antenna, which has unity gain (0 dBi). The power received per MHz is $A_e P_d$. Our UWB signal is not a single frequency, but will have power uniformly spread from some lower frequency f_{low} to an upper frequency f_{up}. We integrate the power density over that band to obtain the total received power

$$P = \frac{(f_{up} - f_{low})c^2}{(4\pi f_{up} f_{low})} \qquad (8.9)$$

Note that the "effective center frequency" of the UWB signal propagation is the geometric mean of the upper and lower band edges.

8.2 The UWB Fundamental Limit

The UWB Fundamental System Gain Limit [Siwiak 2002], SG_{UWB}, is boxed in by the most efficient possible modulation. This is a limit of nature. Regulations delineate the frequency bands and power spectral density limits of the box and, hence, the highest EIRP available to us. We can now combine these into UWB Fundamental System Gain Limits that define the theoretical limit of system gain per bit rate. These values, one in each regulatory flavor, form the basis for UWB system link margin calculations without regard to a specific technology.

Pulses are Boxed in by Limits

8.2.1 Fundamental Limit for UWB

The UWB Fundamental Limit for US regulations is shown in Figure 8.2. The Limit is calculated from the maximum available EIRP and the limit of modulation efficiency relative to the thermal noise floor

$$SG_{lim} = EIRP - N_1 - e_{b\,min} \qquad (8.10)$$

where $N_1 = -174\,dBm$ per Hz. With US regulatory limits, $SG_{lim} = 173\,dBm/bps$. Equation (8.10) states that, in the limit, there is 173 dB of path attenuation available for a 1-bps UWB radio link. Table 8.4 lists the Fundamental Limits for the United States based on the FCC Report and Order [FCC15 2002] and for other jurisdictions where UWB experimentation is permitted, such as Singapore [IDA 2003], or is under active consideration [ETSI 2003].

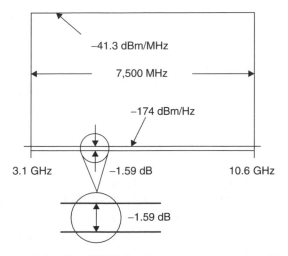

Figure 8.2   The UWB fundamental system gain limit.

Table 8.4   Fundamental UWB limits in various jurisdictions.

Jurisdiction	UWB (GHz)	Emission level (dBm/MHz)	Maximum (EIRP, dBm)	Fundamental limit SG_{UWB} (dB/bps)
US FCC	3.1–10.6	−41.3	−2.55	173.0
Singapore UFZ	2.2–10.6	−35.3	+3.94	179.5
ETSI indoor	3.1–10.6	−41.3	−2.55	173.0
ETSI handheld	3.1–10.6	−61.3	−22.55	153.0

We can now use the Fundamental Limit to calculate the maximum possible system capacity versus distance using free space propagation between 0-dBi gain antennas. In each of the four cases, the link capacity and range follow a straight line for link capacities smaller than the width of the UWB band. In this region, UWB bandwidth and range can be traded off as two octaves of bandwidth per octave of distance – a straight line on the log(distance) versus log(capacity) curves. For capacities higher than the available UWB bandwidth, there is an exponential compression in capacity as distance decreases linearly. That is the price demanded by Shannon's Equation (8.3). Link capacities greater than the available bandwidth can be achieved, but the price is a higher level modulation, which demands an exponentially increasing SNR for a given level of performance.

The path attenuation $P_L(d)$ in free space between two unity gain antennas spaced a distance d apart was given previously by Equation (6.1). The maximum possible received SNR_R in decibels for a data bandwidth B is then

$$SNR_R = SG_{\lim} + P_L(d) - 10 \, \log(B) \qquad (8.11)$$

We can now calculate the UWB Fundamental Limit capacity by inserting (8.11) into (8.3), so that

$$C = B \log_2(1 + 10^{SNR_R}). \qquad (8.12)$$

The resulting capacities C for the Limits in Table 8.3 are shown in Figure 8.3. Note the straight-line relationship between capacity and link distance as long as C does not exceed $B = 7.5\,\text{Gbps}$ (8.4 Gbps, Singapore). Power is consumed at an exponential rate when $C > B$, so the curves flatten out in this region.

Practical UWB radio systems, of course, would be in the neighborhood of 20 dB below the Fundamental Limit in performance because of implementation losses and because not all of the available bandwidth would be used for EIRP. Because of the large available bandwidth, UWB offers a significant advantage over conventional bandwidth-constrained radio technologies. At high data rates, UWB requires far less SNR than conventional approaches, and hence a far smaller emitted power requirement. This is

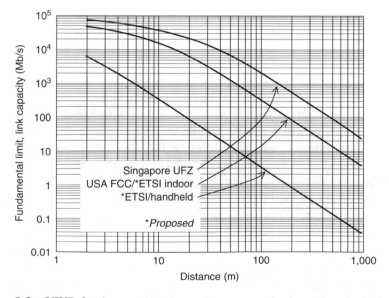

Figure 8.3 UWB fundamental limit maximum capacity in various jurisdictions.

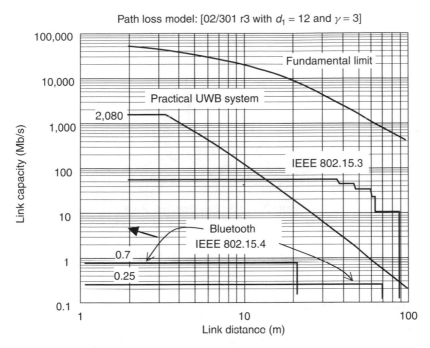

Figure 8.4 UWB and narrowband systems compared.

reflected in Figure 8.4, where a UWB system based on the US limits is compared with 802.15.3, Bluetooth and 802.15.4 systems. All of the 802 systems emit more power than the UWB system. Yet, the UWB system outperforms all of them in the shorter ranges below about 15 m. Some UWB systems can also scale bandwidth for power very easily, so, they can reach out with lower data rates to larger distances, also depicted in Figure 8.4.

In Figure 8.4, the link budget for the "Practical UWB" system is 22-dB worse than the Fundamental limit. In the next section, we will discover the inefficiencies that lead to this discrepancy. Systems like IEEE 802.15.3 are operating in very limited bandwidth and require high SNRs, and therefore a great deal of power. It is, therefore, in the shorter ranges, below about 15 m, that UWB outperforms these higher power systems. Radiated signals do not stop at a particular distance; rather, they become weaker. Some implementations of UWB can easily trade data bandwidth for SNR, so that they can provide greater ranges, but at significantly lower data bandwidth. This, then, is the value proposition for UWB: exceptionally high data rates at the shorter ranges – much higher than is possible with conventional band-limited radio – and low data rates at longer ranges.

Signals Do Not Die They Just Fade Away

8.2.2 Fundamental Limit for Conventional Systems

A Fundamental Limit can be calculated for narrowband radio systems occupying the 2.4-GHz band like IEEE 802.11b and the 5-GHz bands like IEEE 802.11a and HiperLAN/2 [ETSI 2003a] by applying Equation (8.10). The resulting narrowband Limits are in the 190 to 200-dB/bps region, depending on the allowed EIRP and bandwidths available in various jurisdictions. This is significantly higher than the UWB Fundamental Limit calculated for US regulations, but the Limit does not tell the full story. The Limits state that when the systems are compared at the same capacity *and* the available RF bandwidth is larger than the capacity, a larger System Limit signifies better performance. The full story emerges when Equation (8.10) is used with Equations (8.11) and (8.12) to relate capacity to system range. For capacities smaller than the available system bandwidth, the Limiting capacity curve operates in the exponential region of Shannon's formula.

The behavior is evident in Figure 8.5. The hypothetical 2.4- and 5-GHz systems have 80 and a few hundred megahertz of bandwidth available. So, in the longer distances these systems outperform the UWB system. At closer distances, the UWB system is still operating in the linear C to B region of Shannon's formula (Equation 8.12), whereas the narrowband signals require exponentially increasing SNR (more power!) to increase the bandwidth. The difference between the narrowband system limit of 200 dB/bps and UWB's 173 dB/bps is consumed exponentially as the range decreases, and UWB outperforms the narrower band systems at the closer distances.

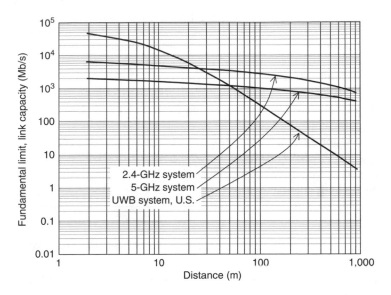

Figure 8.5 UWB and narrowband system fundamental limits.

8.3 UWB Wireless Links

We can estimate the link margin remaining for propagation attenuation by starting with the Fundamental System Gain Limit of 173 dB/bps (in the United States). This is our total budget, which can now be spent on bandwidth, modulation efficiency, actual signal bandwidth occupied, and various implementation losses. We need not concern ourselves at this stage with any specific UWB radio technology. Instead, we specify a channel-link bandwidth, the rate at which we sent bits over the channel. We also choose the modulation employed; hence, we know the modulation efficiency. Finally, we decide on the UWB signal bandwidth, which establishes the total power we can emit. All of these subtract from the fundamental limit of 173 dB/bps available to us.

8.3.1 UWB Link Budgets

Table 8.5 shows a margin of 62.9 dB available to be used up in propagation path losses, which determines the range of our UWB radio link. In Table 8.5, the link can transfer 515 Mbps using an efficient 64-BOK modulation and coding. The UWB signal design has an equivalent power bandwidth of 1.4 GHz. It is now an engineering detail, albeit a nontrivial one, to design a particular UWB system with the design choices shown

Table 8.5 An efficient, high data rate UWB radio system.

System parameter	Design choices	Value
System gain limit, SG_{lim}		173.0 dB/bps
$-10 \log(link\ BW)$	$link\ BW = 515$ Mbps	-87.1 dBbps
$-$Modulation efficiency	64-BOK at 10^{-3} BER	-5.7 dB
System losses, (noise figure, implementation losses)		-10.0 dB
$-10 \log(\text{actual } RF\ BW/7.5\ \text{GHz})$	$RF\ BW = 1.4$ GHz	-7.3 dB
Antenna gain or loss		0 dBi
Available for path attenuation		62.9 dB

in Table 8.5. One such design was introduced as the DS-UWB system when we discussed standards in Chapter 3. There, a rate 0.87 forward error correction (FEC) code was additionally applied, resulting in a net 448-Mbps data throughput rate on this UWB link.

Many other designs are possible, including one that delivers a moderate user throughput of 112 Mbps desired for the IEEE 802.15.3a UWB standard. Here, a 0.44 rate FEC code is desired, so the link bit rate is 256 Mbps. A simple polarity modulation (BPSK) is employed and the UWB signal is designed to have a power bandwidth of 1.5 GHz. Table 8.6 shows that a margin of 66.6 dB is available for propagation path losses. At a center frequency of 4.1 GHz, this margin enables a free space path-link distance of 10 m, with 2.5 dB to spare. In this case, the design choice budget for the receiver noise figure and implementation losses is a challenging 7 dB. Again, the formidable engineering details of implementing this system are omitted here, but the basic parameters have been chosen.

Table 8.6 A medium data rate UWB radio system.

System parameter	Design choice	Value
System gain limit, SG_{lim}		173.0 dB/bps
$-10 \log(link\ BW)$	$Link\ BW = 256$ Mbps	-84.0 dBbps
Modulation efficiency	BPSK at 10^{-3} BER	-8.4 dB
System losses, (noise figure, implementation losses)		-7.0 dB
$-10 \log(\text{actual } RF\ BW/7.5\ \text{GHz})$	$RF\ BW = 1.5$ GHz	-7.0 dB
Antenna gain or loss		0 dBi
Available for path attenuation		66.6 dB

Table 8.7 A low cost, low data rate UWB radio system.

System parameter	Design choice	Value
System gain limit, SG_{lim}		173.0 dB/bps
$-10 \log(link\ BW)$	*Link BW* $= 100$ kbps	-50.0 dBbps
Modulation efficiency	2-TR at 10^{-3} BER	-12.5 dB
System losses, (noise figure, implementation losses)		-13.0 dB
$-10 \log($actual *RF BW*$/7.5$ GHz$)$	*RF BW* $= 1.4$ GHz	-7.3 dB
Antenna gain or loss		0 dBi
Available for path attenuation		90.2 dB

One further set of design choices is shown in Table 8.7. Here, we choose
to target the implementation cost, but we do not have aggressive requirements for performance other than the communications range. The channel
data rate is a modest 100 kbps, and a 2-TR modulation is chosen. This
modulation is not particularly efficient (see Table 8.2) but it is very simple
to implement. With low cost in mind, we are also generous in the noise
figure and implementation-loss budget of 13 dB. We choose a signal design
with a 1.4-GHz power bandwidth centered at 4.1 GHz, resulting in a net
margin of 90.2 dB left for path attenuation. In free space, this would result
in a free space link range of about 220 m. Indoors, this range would be in
the vicinity of 50 m (using the SBY model (Equation 6.20) with $d_t = 3$
and $n = 3$).

These link budgets show the wide range of possibilities for UWB
radio implementations. While devoid of engineering design specifics, the
basic design choices have been made. A wide variety of implementations
is possible.

8.3.2 Receiver Sensitivity and System Gain

Receiver sensitivity is often a very useful parameter in radio system analysis and design. System gain is found when receiver sensitivity is compared
with a transmitter EIRP with receiver antenna gain included. Receiver
sensitivity is the signal level required at the receiver input terminals to
produce a specified level of performance. This performance level is often
a desired BER like, for example, the one listed in Table 8.1. In more
elaborate or complex data protocols, a frame error rate, or some similar
or convenient parameter can be used. EIRP is, of course, the transmitter effective isotropically radiated power. Receiver sensitivity in AWGN,

expressed in decibels, is

$$W = 10 \, \log(N_0 B) + e_b + L_I. \tag{8.13}$$

The noise term is $-174\,\text{dBm}$, B is the receiver system effective noise bandwidth, and e_b is the modulation efficiency (see Table 8.1). L_I includes receiver implementation losses such as noise figure and detector inefficiencies. W is the minimum signal power level at the receiver input terminals that results in the performance level specified for the modulation efficiency. Clearly, receiver sensitivity can be improved by choosing the highest possible modulation efficiency, which means the smallest value of e_b in Tables 8.1 or 8.2.

The system gain is the difference in decibels between the EIRP and the receiver sensitivity added to the receiver antenna gain. System gain S_g is

$$S_g = \text{EIRP} - W + G_{\text{ant}} \tag{8.14}$$

and this is the amount of margin available for propagation losses and interference immunity.

It is interesting to compare the receiver sensitivities and actual system gains for the UWB systems introduced in the previous section. Table 8.8 shows this comparison. Since the maximum available EIRP (for the United States regulations) is $-2.55\,\text{dBm}$, the EIRP in a 1.4- and 1.5-GHz bandwidth are -9.9 and $-9.6\,\text{dBm}$ respectively.

8.3.3 Advantage of UWB in Non-AWGN Channels

Indoor radio channels are subject to multipath because of the many reflecting objects, walls, and clutter present. This affects our link budget in two ways. First, the "strongest impulse" loses energy to wave expansion (free space propagation) and, additionally, sheds energy to multipath dispersion. As shown in Chapter 6, this results in a propagation law closer to the $n = 3$ rather than the $n = 2$ of free space. Second, depending on the

Table 8.8   Receiver sensitivities and system gains compared.

	System of Table 8.5	System of Table 8.6	System of Table 8.7
W	$-72.8\,\text{dBm}$	$-76.2\,\text{dBm}$	$-100.1\,\text{dBm}$
S_g	$62.9\,\text{dB}$	$66.6\,\text{dB}$	$90.2\,\text{dB}$

design of the UWB signal, the multipath dispersion can result in anywhere from no fading to full Rayleigh fading. The statistics of signal fading are summarized in Figure 8.6 for signal bandwidths of 4 MHz, 75 MHz, and 1.4 GHz. The figure shows that a 4-MHz wide signal behaves very much like a Rayleigh-faded signal. Twenty-five percent of the time the 4-MHz wide signal level is faded *more* than 6 dB below the mean signal level. In contrast, a UWB signal that has 1.4-GHz bandwidth fades *less* than 1 dB for 75% of the time.

Clearly, UWB signals have significant advantages over narrowband signal designs. It is important to note, as seen in Chapter 3, that narrowband signals for example, each 4 MHz wide, can be aggregated to occupy a bandwidth greater than 500 MHz. These signals can then access the UWB spectrum under the US regulations. While these are legitimately "UWB" under the regulations, their behavior, as seen in Figure 8.6, can be decidedly narrowband! Aggregating 500 MHz worth of narrowband signals can masquerade as UWB for the purposes of spectrum access, but physics still dictate narrowband behavior in multipath. Figure 8.6 reveals that such signal designs can result in a severe link-margin loss. Figure 8.7 shows the narrowband (4 MHz) and UWB (1.4 GHz) signal probability density functions (PDF) compared.

The PDFs of the narrow and UWB signals clearly show one of the significant advantages of UWB. Wide bandwidth signals do not persist long enough in time to overlap significantly with enough time-dispersed

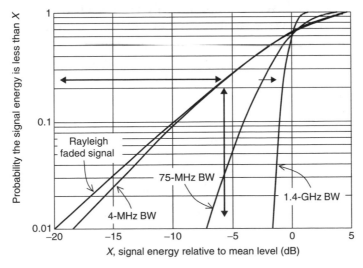

Figure 8.6 Signal statistics for various bandwidths.

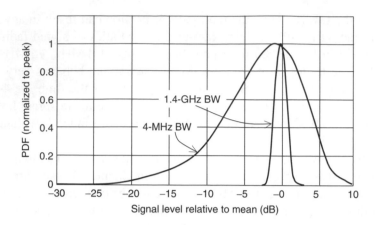

Figure 8.7   Probability densities of narrow and UWB signals.

copies of the signals to fade like narrowband signals. This is shown by the very narrow PDF of the UWB signal compared with the fully faded narrowband signal.

8.4 Link Capacity

The UWB link capacity is simply a measure of how many bits per second can be transferred over a link. In the absence of interference and with no multipath dispersion, this capacity is bounded by Shannon's limit. In practice, however, AWGN is not the only term against which our signal must be discerned. We must also contend with time-delayed copies of our own signal echoes. In a system of multiple users, we must additionally contend with competing energy from the other users. We will develop very simple capacity models for UWB and for narrowband wireless networks. We speak of *ergodic capacity* here, that is, capacity averaged over many realizations.

Pulses Collide in Multipath

8.4.1 A UWB Link in Multipath

In a solitary link involving UWB impulses, the link capacity is related to the number of pulses that can be discerned one from the other. The simplest example is a link sending pulses that are T_b seconds in duration at a rate of $B = 1/T_b$ pulses per second. No multipath is present and the link capacity is then governed by Equation (8.12).

Figure 8.8 shows pulses that are T_b seconds long and spaced by T_b seconds. Any closer together and the pulses would overlap, causing interpulse interference. The pulses fill the bandwidth of the channel.

On an average, the multipath energy (the square of amplitude) profile of pulses in a channel tends to have an exponential energy decay with time as seen in Figure 8.9. (See also the IEEE 802.15.3a channel model CM-4 profile for which 100 random realizations are portrayed in Figure B.4 in Appendix B.) A simple way to think about multipath interference in an ergodic sense is to note that the "echoes" die out exponentially. The next pulse can be discerned only if it is placed beyond where the energy has decayed by at least m_b, where $10 \log(m_b) = e_b$ from Table 8.1. The previously sent pulse-signal energy is the exponentially distributed interference term

$$I(t) = S \exp\left(\frac{-t}{\tau_{RMS}}\right) \tag{8.15}$$

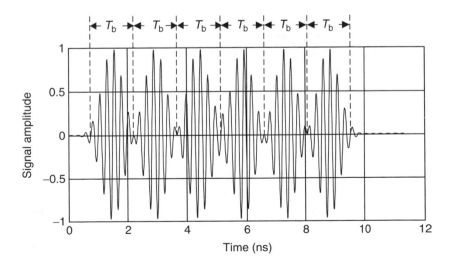

Figure 8.8 Pulses in the absence of multipath.

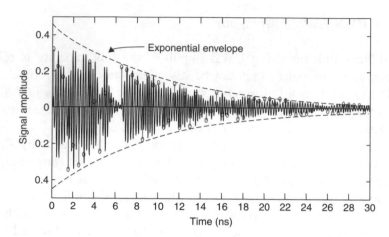

Figure 8.9   Pulse amplitude profile in multipath for a single sent pulse.

where S is the signal energy and τ_{RMS} is the RMS delay spread of the multipath. The signal S to interference plus noise $(I + N)$ ratio must exceed m_b, so

$$\frac{C}{B} = \frac{1}{1 + \dfrac{\tau_{RMS}}{T_b}} \ln(m_b) \qquad (8.16)$$

An important observation about Equation (8.16) is that the capacity decreases as the modulation efficiency decreases (m_b gets bigger). Capacity also decreases as the RMS delay spread increases. This is reflected in Figure 8.10, where the capacity (normalized to the bandwidth) is shown versus the delay spread (normalized by the impulse duration). The most efficient modulations retain the best capacity in multipath.

8.4.2 A Capacity Model for UWB

A capacity model for UWB can now be constructed on the basis of the simple model of the previous section. First, we select some desired system parameters like pulse duration, pulses per symbol, symbol length, and channel attenuation model. We then select the nature of the multipath impairment: the RMS delay spread. Finally, we perform the following algorithm a statistically significant number of times. (Figure 8.11 shows the simulation for a UWB system.)

1. Randomly select "Wanted" link, include random draw for lognormal shadowing.

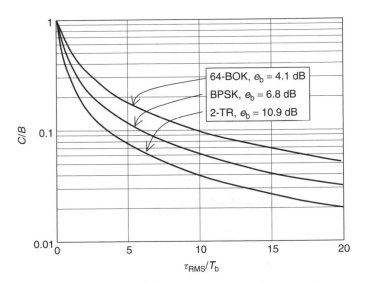

Figure 8.10 Capacity of various modulations in multipath.

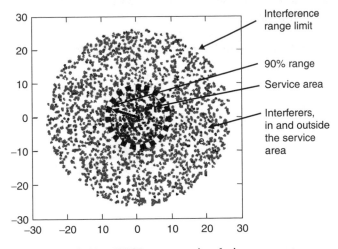

Figure 8.11 UWB system simulation geometry.

2. Randomly select interferers in proportion to the interference area compared with the coverage area.

3. Sum up interference power, include lognormal shadowing and AWGN.

4. Record a "Success" *if*

   "Wanted energy"/"Interference energy" > "needed $S/(I + N)$."

5. Repeat the experiment a sufficient number of repetitions ($R =$ 150,000+).

6. Talley up the "Success" factor F as the fraction of "Success" outcomes.

The UWB system link capacity is then calculated as the maximum capacity without multipath multiplied by the factor F. The resulting value is the shared system capacity as averaged over the coverage area. It is an indication of system performance in an environment containing many same-system users.

8.4.3 Capacity Model for IEEE 802.11a

A similar statistical experiment is constructed for the IEEE 802.11a wireless system. None of the medium access control (MAC) layer functions are modeled, and one data rate at a time is considered. A MAC efficiency of 0.60 is assumed. The intent is to keep the model simple and comparable to the UWB model. In this simple model, 8 frequencies are available and are arranged in a 4-cell reuse pattern as seen in Figure 8.12. The 802.11a algorithm is as follows:

1. Randomly select "Wanted" link, include random draw for lognormal shadowing.

2. Randomly select 4 on-channel, 2 adjacent, and 2 alternate-adjacent channel interferers.

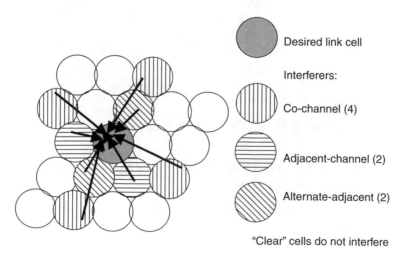

Figure 8.12   System layout for the 802.11a capacity simulation.

3. Sum up interference power, include lognormal shadowing and AWGN.

4. Record a "Success" *if*

   "Wanted energy"/"Interference energy" > "needed $S/(I + N)$."

5. Repeat the experiment a sufficient number of repetitions ($R = 150,000+$).

6. Talley up the "Success" factor F as the fraction of "Success" outcomes.

The 802.11a system link capacity is then calculated as the maximum capacity without multipath, including all available channels in the band, divided by the cell-reuse factor, and multiplied by the factor F. The resulting value is the shared system capacity as averaged over the coverage area. It is an indication of system performance in an environment containing many same-system users.

8.4.4 Capacity Model for IEEE 802.11b

An IEEE 802.11b wireless system model is constructed along the same lines. Again, none of the MAC layer functions are modeled, and one data rate at a time is considered. An MAC efficiency of 0.67 is assumed. The 802.11b system has three frequencies available and they are arranged in a three-cell system as shown in Figure 8.13. The intent is to keep the model simple and comparable to the UWB model. The 802.11b algorithm is as follows:

1. Randomly select "Wanted" link, include random draw for lognormal shadowing.

2. Randomly select an average of adjacent channel interferers.

3. Sum up interference power, include lognormal shadowing and AWGN.

4. Record a "Success" *if*

   "Wanted energy"/"Interference energy" > "needed $S/(I + N)$."

5. Repeat the experiment a sufficient number of repetitions ($R = 150,000+$).

6. Talley up the "Success" factor F as the fraction of "Success" outcomes.

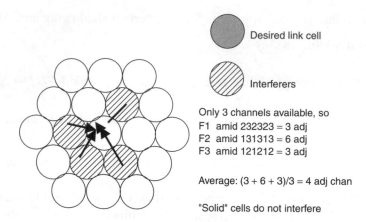

Figure 8.13   System layout for the 802.11b capacity simulation.

The 802.11b performance, adjacent channel protection and other system parameters are taken from the 802.11b specification [IEEE802.11b]. The 802.11b system link capacity is then calculated as the maximum capacity without multipath, including all available channels in the band, divided by the cell reuse factor, and multiplied by the factor F. The resulting figure is the shared system capacity as averaged over the coverage area. It is an indication of system performance in an environment containing many same system users.

8.4.5 Comparing UWB with the 802s

The very simple models for UWB capacity and 802.11a/b capacities were chosen for comparison in a 14 ns RMS delay spread environment. Capacity analyses can be made more elaborate and more complicated, and these would tend to yield refined answers. However, they would not give substantially different results from the ones presented in Table 8.9. Measurements of capacity [Atheros 2001], for example, which included the MAC operation, compared the 802.11a and 802.11b systems, and gave results similar to this analysis. The conclusion from the measurements was that the 11a capacity ranges from 17- to 26-Mbps throughput depending on the range, while the 11b system capacity will be around 5 Mbps. Our simple analysis shows that the average cell shared capacity, which is comparable to throughput, peaks at 20 Mbps when one system data rate at a time is considered. The simple analysis reveals further details that did not directly surface from measurements. The single-link (one pair of users) realized capacity increases as the system data rate is increased. The

Table 8.9   Comparison of UWB and IEEE 802.11a and 11b system capacities.

Technology	Single-link data rate, (Mbps)	Single-link realized, capacity, (Mbps)	Average cell shared capacity, (Mbps)
802.11a (8 channels, 4 cell reuse)	6	3.6	6.3
	12	7.2	12.2
	24	14.4	20.0
	36	21.6	19.0
	54	32.4	12.5
802.11b (3 channels, 3 cell reuse)	1	0.7	0.7
	2	1.3	1.3
	5.5	3.6	3.6
	11	7.3	7.3
UWB in 1.4 GHz BW	256	112	83.0

802.11a system shared capacity (throughput), however, peaks at 20 Mbps and then decreases as higher system data rates are chosen! At the highest data rate of 54 Mbps, the average shared-cell capacity is significantly smaller than at the 24-Mbps and 36-Mbps data rates! This should not come as a surprise, since the higher data rates in 11a rely on a very fragile high-level OFDM modulation because the bandwidth is restricted. This is a subtle advantage of UWB, which uses the most robust modulation at all data rates.

The 802.11b system does not experience as much adjacent interference as 11a, because the channels are well separated and the modulation is a robust direct sequence spread spectrum (DSSS). Single-link capacity equals the average shared-cell capacity. The UWB system of Table 8.6 starts with a link rate of 256 Mbps and a single-link realized capacity of 112 Mbps. Multipath and self-system interference reduces that to a still respectable 83 Mbps shared capacity.

8.5 Summary

UWB, like any radio system, does not operate in a perfect world. Both thermal noise and human-caused interference limits wireless system performance. A theoretical maximum limit on UWB "system gain" per bit per second of 173 dB/bps provides us with a simple means by which to estimate the performance and capabilities of practical radio links. In particular, UWB links can be easily analyzed without specifically resorting to the technological detail.

Link budgets can also be easily determined. We found that system capacities can be determined with relatively simple models that allowed UWB and 802.11a and 11b systems to be compared. The analysis yielded the confirmation that fragile modulations, as are needed for the highest data rates in a bandwidth-starved system, fall apart easily in scenarios of multiple users.

References

[Atheros 2003] Atheros White Paper, 802.11 Wireless LAN Performance, (Online): <http://www.atheros.com/pt/atheros_range_white-paper.pdf>, 14 December 2003.

[CODATA 2002] *2002 CODATA Recommended Values of the Fundamental Physics Constants*, 31 December 2002, (Online): <http://physics.nist.gov/cuu/Constants/>, December 2003.

[Couch 1993] L. W. Couch, *Digital and Analog Communication Systems*, New York: Macmillan Publishing, 1993.

[FCC15 2002] US 47 CFR Part15 Ultra-Wideband Operations FCC Report and Order, 22 April 2002.

[IDA 2003] IDA (Infocomm Development Authority of Singapore), *Singapore Ultra-Wideband Programme*, Singapore 038988.

[IEEE802 03/105] IEEE802 Document: <03105r1P802-15_TG3a-General-Atomics-CFP-Presentation.ppt>, 7 March 2003.

[IEEE802.11a] IEEE Std 802.11a-1999 (Supplement to IEEE Std 802.11–1999) [Adopted as ISO/IEC 8802-11:1999/Amd 1:2000(E)], *Part 11: Wireless LAN Medium Access Control (MAC) and Physical Layer (PHY) Specifications High-Speed Physical Layer in the 5 GHz Band*.

[IEEE802.11b] IEEE Std 802.11b/D8.0, September 2001, (Draft Supplement to IEEE Std 802.11 1999 Edition), *Part 11: Wireless LAN Medium Access Control (MAC) and Physical Layer (PHY) Specifications: Higher Speed Physical Layer (PHY) Extension in the 2.4 GHz Band*.

[ETSI 2003] ETSI, *Harmonised Standards Covering Ultrawide Band (UWB) Applications*, Directorate General of the European Commission, Standardisation Mandate: DG ENTR/G/3M/ 329, Brussels, 25 February 2003.

[ETSI 2003a] HiperLAN/2 Standard for Broadband Radio Access Networks, (Online): <http://portal.etsi.org/bran/kta/Hiperlan/hiperlan2. asp>, 14 December 2003.

[Lindsey 1973] W. C. Lindsey and M. K. Simon, *Telecommunication System Engineering*, New York: Dover Publications, 1973.

[McKeown 2003] D. McKeown, *Gammz UWB Cartoons and Art*, Private Communication to K. Siwiak, December 2003.

[Schantz 2003] H. G. Schantz, *Introduction to Ultra-Wideband Antennas*, IEEE UWBST 2003, Reston, Virginia, 16–19 November 2003.

[Shannon 1948] [C. E. Shannon, *"A mathematical theory of communication"*, *The Bell System Technical Journal*, **27**, 379–423, 623–656, 1948.

[Siwiak 2002] K. Siwiak, "UWB and 173", *WCNC 2002*, Orlando, FL, March 17–21 2002.

9

Applications and Future Directions

Introduction

Progress and innovation occur fastest and evolve in unexpected ways when two or more processes converge in time. UWB impulse technology has been around for many decades. Digital logic has also been around for many decades. The convergence of wideband impulse technology with the decreasing cost and increasing capabilities of digital logic enabled the emergence of UWB as an appealing technology. It became economical enough to build impulse radios that could offer precision distance measurements, high data rate communications, and high-resolution radars for nongovernment markets. The fusion of all of these capabilities has already been demonstrated in early UWB prototypes [Siwiak 2002]. This emergence of commercially viable impulse radio drove and converged with regulatory evolution that now permits UWB to appear in a large tract of shared spectrum.

The regulatory process too has been evolving. We saw that the earliest regulations fostered the growth of narrowband radio technologies because that was the right answer at the time for the economic growth of radio. Changes to that narrowband mandate happened when other technologies matured to a point of economic viability. Narrowband-channelized radio produced commercial AM broadcasting, which fueled a need for high-quality audio programming. Blend that with wideband FM technology and we have "high-fidelity" FM broadcasting. This, in turn, nudged regulations

Ultra-Wideband Radio Technology   Kazimierz Siwiak and Debra McKeown
© 2004 John Wiley & Sons, Ltd   ISBN: 0-470-85931-8

away from the strict, early narrowband vision. Theoretical advances in communications theory coupled with the increasingly economical digital processing capabilities gave rise to spread-spectrum technology. Again, the regulations accommodate progress and spectrum was set aside for the technology. We saw the same with UWB. Regulations permitting the technology are now in place in the United States and are being contemplated elsewhere. However, the broadly written regulations converged with the standardization process, which produced unexpected results in UWB radio design proposals. The convergence enabled conventional narrowband techniques to masquerade as UWB under the provisions of the rules. These now emerge as viable paths to a perceived, vast, commercial market place for high data rate communications.

9.1 A Time Line of Wireless

We pause to gaze into our crystal ball now, to see the past unfurl into the present, with the promise of tomorrow's trends and capabilities unfolding before our eyes. Technology, marketing, and consumer needs along with radio regulations, interact closely throughout history, as seen in Figure 9.1. Radio was invented in a crude and coarse form. Immediate and important safety-of-life applications like ship-to-shore radio led to the need for better spectrum usage. At the time, that meant cleaning up the signals and mandating that they each utilize as little of the precious spectrum as possible. The regulatory focus was on *strict frequency management*. When radio

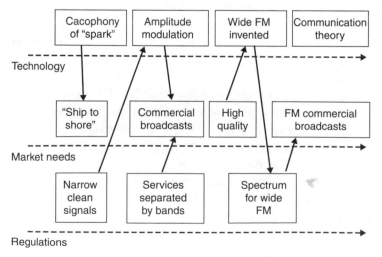

Figure 9.1 The interplay of technology and market needs with regulations.

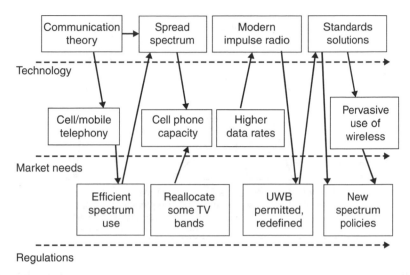

Figure 9.2   Market needs driven by innovation lead to pervasive use of wireless.

became capable enough to sustain commercial broadcasting, the need for higher quality audio broadcasting was answered by technology as well as by evolved regulations.

Regulations did allow for wider band technologies when there was a clear marketing and economic need. AM gave way to FM commercial broadcasting, as seen in Figure 9.1. All along, communications theory developed and fueled innovations while technologies converged to give us economical cell and mobile telephones (Figure 9.2). Increased mobile phone use feeds the need for more data and at higher data rates. Wireless devices are pervading our society. Progress in the Standards definition process has also converged with UWB and UWB regulations. Standards activities [IEEE802 2003] are promoting a global perspective of not only technology but also of regulations, as shown in Figure 9.3.

Globally, regulators are watching the Standards process to help guide them in the way the radio spectrum will be managed in the future. We saw recent as well as current trends to reassign blocks of spectrum from television service usage to mobile phone and two-way radio usage. Now we have regulations that permit UWB in response to the emergence of commercially viable UWB technologies. However, the regulations also changed the character of UWB. Once thought of in terms like "impulse radio" and multigigahertz bandwidths, UWB is now defined in terms that are broad enough to allow variations of conventional narrowband technologies to appear. UWB is flexible enough in its definitions,

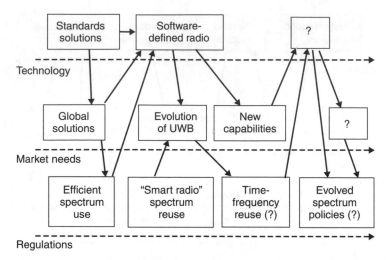

Figure 9.3   Spectrum management evolves to accommodate newer wireless capabilities.

implementations, capabilities, and regulations that it may evolve to the "smart radios" of the near future. Already regulators in the United States are contemplating the next phase of spectrum management (see [FCC 2003]), whereby spectrum is to be reused and reallocated on the spot in response to local spectral sensing and local needs. UWB is poised for this inevitable evolution.

UWB is Flexible

9.2 UWB Applications

Ultra-wideband is the contortionist of the wireless world – it is flexible enough to work in many different ways while still maintaining its character. Because it can be used with such diversity, there are almost infinite manners in which to utilize UWB in applications that run the gamut from invaluable to ingenious to ridiculous. These applications are distributed amongst three categories:

- Communications and sensors
- Position location and tracking
- Radar.

UWB has characteristic attributes that make it especially attractive in those three categories. These include the following:

- *Stealth* – in many varieties of UWB, the signal appears to be very low-level background noise to an unintended narrowband receiver.
- *Local area networks* – the exceptionally large available bandwidth can be used as the basis for a short-range wireless local area network with data rates approaching gigabits per second.
- *Position location* – some UWB systems are capable of determining the 3D location of any of its transponders to within a few centimeters.
- *Radar imaging* – UWB systems can be used as an open-air through-wall or ground-penetrating radar imager.
- *Security bubble* – some UWB systems can be configured to create a security "bubble" in the fashion of a bistatic radar to detect penetration of the bubble walls.
- *Vehicular radar systems* – UWB has an available vehicular radar band in the frequency range 22 to 29 GHz for use in collision avoidance and parking aids.

Below are a few of the ideas that have been considered, most of which were chosen for presentation here because of their market viability. However, one only needs to use a bit of imagination and countless applications will spring into mind.

9.2.1 Communications and Sensors

Applications for communications present some of the most exciting opportunities in the consumer market. Communication is a part of our daily life and the ways in which it can be enhanced, enriched, and made more efficient with UWB are endless. Applications in communications can be classified into two areas – low or high data rates. Both require low power and high capacity, which are the star qualities of UWB.

Low Data Rate   Low data rate devices surround us in our technological world – but they are usually attached by wires and cables. We use these items to enter data into or retrieve data from our computers, to detect home intruders, and for countless other purposes. Low data rate devices can effectively be wireless, but the solutions on the market today are bound by line-of-sight interference with other devices, power issues, and other less than ideal compromises. UWB is not constrained by line of sight quite as dramatically as is infrared light, since the wavelengths are long by comparison and can generally bend around or transmit through objects without impeding the connection. It is also less affected by shadows and other light-related interferences than is the case with infrared. Since UWB operates at such low power and in intermittent fashion, interference is not significant either – that means that hundreds of devices could operate in the same space without intruding on each other.

Given the numerous benefits UWB offers over current technologies on the market, let us consider some of the applications that could be improved or created using this innovative method of communicating at low data rates. We will first consider existing items whose performance could be greatly enhanced by using UWB. Consider a garage door opener that would open only your garage and never your neighbors. Special capabilities of UWB, like knowing the distance to the garage door, can create more security for you and your neighbors.

Computer peripherals offer another fitting use of UWB, especially when mobility is important and numerous wireless devices are utilized in a shared space. One scenario might include a mouse, keyboard, printer, monitor, audio speakers, microphone, joystick, and PDA – all wireless, all sending messages to the same computer from anywhere in the given range. UWB has the inherent capacity so that this multitude of closely spaced devices can operate and not interfere with each other or with additional computers, which might have an equal number of wireless devices operating at the same time in the same space. This concept is a sure win in the marketplace as it is gratifying across the senses – free movement, community minded, aesthetically pleasing with no wires, and technologically elegant.

Sensor Networks   Sensors of all types offer another opportunity for UWB to flourish. Sensors are currently being used copiously in applications. A variety of sensors are used to secure homes, automobiles, and other property. Installation of modern security systems is time consuming and expensive, though. Why? This is because wires are expensive and time-intensive to install. Often, families cut corners, placing these wired sensors only on the most visible entrances. With a wireless solution, though, the cost of installation and maintenance could drop dramatically and the coverage could be expanded and made more reliable. UWB can be used as the communications link in a sensor network [IEEE802 2003a], and the UWB signal itself can function as the sensor. It could even be tailor-made to form security bubbles around a given area in need of protection, including varying zones of alarm [Time 2003]. Imagine what else could be done with this concept to provide safety, security, and peace of mind. Robert Frost wrote in *Mending Walls*, "Good fences make good neighbors." The best fences are unobtrusive and invisible: the domain of UWB.

Sensors are also being used in medical situations to determine pulse rate, temperature, and other critical life signs. Today, a patient is shackled by wires and cables when extensive medical monitoring is required, as pictured in Figure 9.4. Again, UWB can be used to transport the sensor information without wires, but can also function as a sensor of respiration, heart beat, and, in some instances, for medical imaging.

A UWB sensor network frees the patient from the tangle of wired sensors. Noticeably devoid of wires in Figure 9.5, the UWB solution provides a comfortable "bedside manner" for a patient in need of constant monitoring.

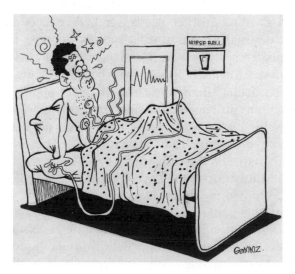

Figure 9.4   Medical sensor wiring can be invasive [McKeown 2003].

Figure 9.5   Removing the sensor wires improves patient comfort [McKeown 2003].

High Data Rate   As the available bandwidth to users expands, applications will continue to evolve and fill the available bandwidth as demand rises. On top of this mounting interest for bandwidth, the increase in mobile telephony and travel has spurred demand for bandwidth mobility, implying wireless technology. The earliest applications of UWB will revolve around existing market needs for higher-speed data transmission. However, demand for multimedia-capable wireless is already driving multiple initiatives in the wireless standards bodies. UWB solutions will emerge tailored for these applications because of the available high bandwidth. In particular, high-density multimedia applications, such as multimedia streaming in "hot spots" like airports or shopping centers or even in multidwelling units, will require bandwidths not currently enabled by continuous-wave "narrowband" technologies. The ability to tightly pack high bandwidth UWB "cells" into these areas without degrading performance will further drive the development of UWB solutions. Downloading of video movie purchase or rental, for example, is a very data-intensive activity that could be enabled by UWB (see Figure 9.6).

Large high-resolution video screens are rapidly becoming available at affordable prices. These devices can benefit from the high capacity capabilities of UWB to stream video content wirelessly from the video source to a wall-mounted screen (see Figure 9.7).

Figure 9.6   Video downloading is very data intensive [McKeown 2003].

Figure 9.7   A video source to a wall-mounted screen is a high data rate application [McKeown 2003].

9.2.2 Position Location and Tracking

Location and tracking on a large scale, for example with GPS, has changed the way we travel. Location and tracking with a smaller range could change the way we organize and track items. These applications could improve security of material goods, help us find our car keys and even keep pace with our children when they are away from us.

Position Location   Today, there are technologies that allow us to pinpoint our location on a globe with accuracy, which has previously been impossible. We have gone from a compass and map to GPS. Now, imagine being able to take that ability one step further – to the indoors. Though UWB is not an efficient solution for outdoor location (the ranges are too short), it is an excellent solution for short-range problems. Some varieties of UWB can be used to determine the range between UWB radios indoors. UWB localizers can be strategically placed in a network of wireless signposts along a trail to mark the route. They can be used to find people in a variety of situations, including fire fighters in a burning building, police officers in distress, an injured skier on a ski slope, hikers injured in a remote area, or children lost in the mall or amusement park [Aetherwire 2003].

Tracking   With advanced tracking mechanisms, we could not only know item locations but actually follow their movement through any number of events. For example, items stored in a warehouse could be tracked from arrival to departure and even to their final destinations. Any in-house movement could be followed as well. Tracking goods could markedly improve our ability to streamline storage and delivery of goods and services while increasing inventory control.

As the mobility of people and objects increases, up-to-date and precise information about their location becomes a relevant market need. While GPS and some E911 technologies promise to deliver some level of accuracy outdoors, current indoor tracking technologies remain relatively scarce and have accuracies on the order of 3 to 10 m. UWB implementations are an adjunct to GPS and E911 that allow the precise determination of location and the tracking of moving objects within an indoor space to an accuracy of several centimeters. This, in turn, enables the delivery of location-specific content and information to individuals on the move, and the tracking of high-value assets for security and efficient utilization. While this is an emerging market segment, the accuracy provided by UWB will accelerate market growth and the development of new applications in this area. UWB systems can work in complex environments where there are many people, assets, and interactions. Places such as hospitals, secure sites, training centers, and distributed workplaces can benefit from faster and more effective communication between people. Automated record keeping for complex unstructured activities can free people from administrative tasks. Equipment profiles can be personalized, automatically enabling sharing of equipment, so that people can get more out of the existing asset base. Real-time measurement and audits of workplace metrics can provide managers with information needed to make practical decisions. Unparalleled levels of security can be achieved by monitoring the locations of people and critical assets [Ubisense 2003].

9.2.3 Radar

UWB signals enable inexpensive high definition radar. With the new radar capability created by the addition of UWB, the radar market will grow dramatically and radar will be used in areas currently unthinkable. Some of the key new radar applications in which UWB is likely to have a strong impact include automotive sensors, collision avoidance sensors, smart airbags, intelligent highway initiatives, personal security sensors, precision surveying, and through-the-wall public safety applications. Radar-enhanced security domes based on precision radar have already demonstrated the capability to detect motion near protected areas, such as high-value assets, personnel, or restricted areas. The dome is software configurable to detect movement passing through the edge of the dome, but can disregard movement within or beyond the dome edge [Time 2003].

Through-wall motion detection capability is a reality today [Radar 2003]. The device sends millions of ultra-wideband pulses per second, creating a signal that, in most circumstances, can penetrate most common building materials, including reinforced concrete, concrete block, Sheetrock, brick, wood, plastic, tile, and fiberglass. The result is an entirely new means of threat detection with many new uses [Withington 2003]. This radar device is targeted towards military and law enforcement tactical teams, for whom detailed knowledge can improve both the effectiveness and safety of a tactical entry.

Operation of vehicular radar in the 22- to 29-GHz band is permitted under the UWB rules using directional antennas on automobiles. These devices are able to detect the location and movement of objects near a vehicle, enabling features such as near collision avoidance, improved air bag activation, and suspension systems that better respond to road conditions.

9.3 UWB Over Wires

UWB technology can be delivered over wire lines and cables. This could effectively double the bandwidth available to cable television (CATV) systems without modification to the existing infrastructure [Pulselink 2003]. Over-wire technology for coaxial cable can provide up to 1.2 Gbps downstream and up to 480 Mbps upstream of additional bandwidth, at low cost,

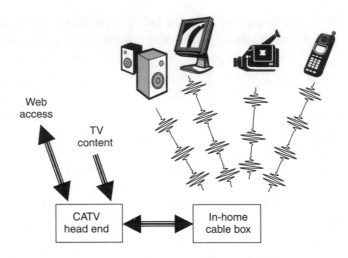

Figure 9.8   Integrated wire-line and wireless UWB technology.

on differing CATV architectures. The UWB signals can be introduced at the cable head end and extracted at the customer's premises. The wire-line UWB technology does not interfere with or degrade television, high-speed Internet, voice or other services already provided by the CATV infrastructure. This will give operators the ability to leverage existing infrastructure to deliver greater functionality in the pursuit of additional revenues. As shown in Figure 9.8, the system uses innovative techniques to seamlessly integrate the UWB-wireless and UWB-wired communications.

This integrated wire-line and wireless technology provides tremendous wireless networking bandwidth that also extends content security all the way from the cable provider's head-end offices out to a variety of wirelessly networked devices. The concept turns the home entertainment center into a wireless hub and networking gateway.

9.4 Summary

UWB operates at ultralow power, transmitting signals over multiple gigahertz, or as little as 500 MHz of bandwidth. UWB operates with emission levels commensurate with levels of unintentional emissions from common digital devices such as laptop computers and pocket calculators. Today, we have a "spectrum drought," in which there is a finite amount of available spectrum, yet there is a rapidly increasing demand for spectrum to accommodate new commercial wireless services. Even the defense community continues to find itself defending its spectrum allocations from the competing demands of commercial users and other government users. UWB exhibits incredible spectral efficiency by virtue of wide bandwidth spreading. UWB technology represents a win–win innovation that makes available efficient use and reuse of critical spectrum to government, public safety, and commercial users. The best applications for UWB are for indoor use in high clutter environments. UWB products for the commercial market will make use of recent technological advancements in transceiver design and will operate at very low power consumption. UWB technology enables not only communications devices but also positioning capabilities of exceptional performance. The fusion of positioning and data capabilities in a single technology opens the door to exciting and new technological developments.

References

[Aetherwire 2003] Aether Wire & Location, Inc., (Online): <http://www. aetherwire.com/>, 7 December 2003.

[IEEE802 2003] IEEE P802.15.3 High Rate (HR) Task Group (TG3) for Wireless Personal Area Networks (WPANs), (Online): <http:// grouper.ieee.org/groups/802/15/pub/TG3.html>, 1 December 2003.

[IEEE802 2003a] IEEE 802.15 WPAN^{TM} Task Group 4 (TG4), <http://www. ieee802.org/15/pub/TG4.html>, 17 December 2003.

[FCC 2003] *FCC Opens Proceedings on Smart Radios*, (Online): <http:// hraunfoss.fcc.gov/edocs_public/attachmatch/DOC-242312A1. pdf>, 17 December 2003.

[McKeown 2003] D. McKeown, *Gammz UWB Cartoons and Art*, Private Communication to K. Siwiak, December 2003.

[Pulselink 2003] Pulse~LINK Inc., (Online): <http://www.pulselink.net/>, 18 December 2003.

[Radar 2003] Time Domain Corp., (Online): <http://www.radarvision. com>, 18 December 2003.

[Siwiak 2002] K. Siwiak and L. L. Huckabee, "Ultra wideband radio", in J. G. Proakis (Ed.), *Encyclopedia of Telecommunications*, New York: John Wiley & Sons, 2002.

[Siwiak 2002a] K. Siwiak and D. McKeown Culver, *UWB Radio Technology in Wireless PANs*, A Tutorial, Birmingham, AL, (Online): <http://www.ieee.org/organizations/eab/icet/presentations. htm>, 7 May 2002.

[Time 2003] Time Domain Corporation, (Online): <http://www.time domain.com>, 17 December 2003.

[Ubisense 2003] Ubisense Limited, (Online): http://www.ubisense.net/index. html>, 17 December 2003.

[Withington 2003] P. Withington, H. Fluhler and S. Nag, "Enhancing homeland security with advanced UWB sensors", *IEEE Microwave Magazine*, **4**(3), 2003.

Appendix A

Excerpts from the *FCC First Report and Order*

This Appendix contains the excerpts from the *Introduction*, *Executive Summary*, and *Appendix D* (changes to the regulations) of the *FCC First Report and Order, ET Docket 98–153* adopted February 14, 2002 and released April 22, 2002 by the Federal Communications Commission (FCC). The Report and Order establishes the basis and justification for the FCC rules; the reader is additionally urged to consult the latest Part 15 rules.

I. Introduction

1. By this action, we are amending Part 15 of our rules to permit the marketing and operation of certain types of new products incorporating ultra-wideband ("UWB") technology. UWB devices operate by employing very narrow or short duration pulses that result in very large or wideband transmission bandwidths. UWB technology holds great promise for a vast array of new applications that we believe will provide significant benefits for public safety, businesses and consumers. With appropriate technical standards, UWB devices can operate using spectrum occupied by existing radio services without causing interference, thereby permitting scarce spectrum resources to be used more efficiently. This First Report and Order ("Order") includes standards designed to ensure that existing and planned radio services, particularly safety services, are adequately protected. We are proceeding cautiously in authorizing UWB technology, based in large measure on standards that the National Telecommunications

Ultra-Wideband Radio Technology   Kazimierz Siwiak and Debra McKeown
© 2004 John Wiley & Sons, Ltd   ISBN: 0-470-85931-8

and Information Administration ("NTIA") found to be are necessary to protect against interference to vital federal government operations. These UWB standards will apply to UWB devices operating in shared or in non-government frequency bands, including UWB devices operated by U.S. Government agencies in such bands. We are concerned, however, that the standards we are adopting may be overprotective and could unnecessarily constrain the development of UWB technology. Accordingly, within the next six to twelve months we intend to review the standards for UWB devices and issue a further rule making to explore more flexible technical standards and to address the operation of additional types of UWB operations and technology.

2. This has been an unusually controversial proceeding involving a variety of UWB advocates and opponents. These parties have been unable to agree on the emission levels necessary to protect Government-operated, safety-of-life and commercial radio systems from harmful interference. It is our belief that the standards contained in this Order are extremely conservative. These standards may change in the future as we continue to collect data regarding UWB operations. The analyses and technical standards contained in this Order are unique to this proceeding and will not be considered as a basis for determining or revising standards for other radio frequency devices, including other Part 15 devices.

3. The following text first provides an executive summary of the major actions taken in this item. Next, a background section describing Part 15 of the Commission's rules and the history of this proceeding is provided. A comprehensive discussion section consisting of several parts is also included. The first section of the discussion focuses on regulatory treatment and the Commission's definition of ultra wideband technology. The next part of the discussion provides analyses of studies submitted by several parties assessing the interference potential of ultra wideband devices to existing services. This section is followed by a discussion of the emission limits established for ultra wideband deployment. Also included in the discussion section are assessments of the cumulative impact of ultra wideband devices and procedures for measuring the emissions from ultra wideband devices. Finally, the discussion concludes with a section on other matters that impact the authorization of UWB technology.

II. Executive Summary

4. Upon consideration of the record, we continue to believe that UWB technology offers significant benefits for Government, public safety, businesses

and consumers. However, we recognize that these substantial benefits could be outweighed if UWB devices were to cause interference to licensed services and other important radio operations. Our analysis of the record and the various technical studies submitted indicate that UWB devices can be permitted to operate on an unlicensed basis without causing harmful interference provided appropriate technical standards and operational restrictions are applied to their use.

5. To ensure that UWB devices do not cause harmful interference, this Order establishes different technical standards and operating restrictions for three types of UWB devices based on their potential to cause interference. These three types of UWB devices are:

1) imaging systems including Ground Penetrating Radars (GPRs) and wall, through-wall, surveillance, and medical imaging devices,

2) vehicular radar systems, and

3) communications and measurement systems.

Generally, we are adopting unwanted emission limits for UWB devices that are significantly more stringent than those imposed on other Part 15 devices; limiting outdoor use of UWB devices to imaging systems, vehicular radar systems and hand held devices; and, limiting the frequency band within which certain UWB products will be permitted to operate. The frequency band of operation is based on the $-10\,dB$ bandwidth of the UWB emission. This combination of technical standards and operational restrictions will ensure that UWB devices coexist with the authorized radio services without the risk of harmful interference while we gain experience with this new technology. In the meantime, we plan to expedite enforcement action for any UWB products found to be in violation of the rules we are adopting and will act promptly to eliminate any reported harmful interference from UWB devices. Specifically, the Order takes the following actions:

- Imaging Systems: Provides for the operation of GPRs and other imaging devices under Part 15 of the Commission's rules subject to certain frequency and power limitations. All imaging systems are subject to coordination with NTIA through the FCC. NTIA has indicated that coordination will be as expeditious as possible, requiring no longer than 15 business days, and may be expedited in emergency situations. The operators of imaging devices must be eligible for licensing under Part 90 of our rules, except that medical imaging devices may be operated by a licensed health care practitioner. Imaging systems include:

- Ground Penetrating Radar Systems: GPRs must be operated below 960 MHz or in the frequency band 3.1–10.6 GHz. GPRs operate only when in contact with, or within close proximity of, the ground for the purpose of detecting or obtaining the images of buried objects. The energy from the GPR is intentionally directed down into the ground for this purpose. Operation is restricted to law enforcement, fire and rescue organizations, to Federal Communications Commission FCC 02–48 scientific research institutions, to commercial mining companies, and to construction companies.

- Wall Imaging Systems: Wall imaging systems must be operated below 960 MHz or in the frequency band 3.1–10.6 GHz. Wall-imaging systems are designed to detect the location of objects contained within a "wall," such as a concrete structure, the side of a bridge, or the wall of a mine. Operation is restricted to law enforcement, fire and rescue organizations, to scientific research institutions, to commercial mining companies, and to construction companies.

- Through-wall Imaging Systems: These systems must be operated below 960 MHz or in the frequency band 1.99–10.6 GHz. Through-wall imaging systems detect the location or movement of persons or objects that are located on the other side of a structure such as a wall. Operation is limited to law enforcement, fire and rescue organizations.

- Surveillance Systems: Although technically these devices are not imaging systems, for regulatory purposes they will be treated in the same way as through-wall imaging systems used by police, fire and rescue organizations and will be permitted to operate in the frequency band 1.99–10.6 GHz. Surveillance systems operate as "security fences" by establishing a stationary RF perimeter field and detecting the intrusion of persons or objects in that field. Operation is limited to law enforcement, fire and rescue organizations, to public utilities and to industrial entities.

- Medical Systems: These devices must be operated in the frequency band 3.1–10.6 GHz. A medical imaging system may be used for a variety of health applications to "see" inside the body of a person or animal. Operation must be at the direction of, or under the supervision of, a licensed health care practitioner.

- Vehicular Radar Systems: Provides for the operation of vehicular radar in the 22–29 GHz band using directional antennas on terrestrial transportation vehicles provided the center frequency of the emission and the frequency at which the highest radiated emission occurs are greater than 24.075 GHz. These devices are able to detect the location and movement of objects near a vehicle, enabling features such as near collision avoidance, improved airbag activation, and suspension systems that better respond to road conditions. Attenuation of the emissions below 24 GHz is required above the horizontal plane in order to protect space borne passive sensors operating in the 23.6–24.0 GHz band.

- Communications and Measurement Systems: Provides for use of a wide variety of other UWB devices, such as high-speed home and business networking devices as well as storage tank measurement devices under Part 15 of the Commission's rules subject to certain frequency and power limitations. The devices must operate in the frequency band 3.1–10.6 GHz. The equipment must be designed to ensure that operation can only occur indoors or it must consist of hand held devices that may be employed for such activities as peer-to-peer operation.

FCC 02–48 Appendix D – Changes to the Regulations

Title 47 of the Code of Federal Regulations, Part 15, is amended as follows:

1. The authority citation for Part 15 continues to read as follows:
AUTHORITY: 47 U.S.C.154, 302, 303, 304, 307 and 544A.

2. Section 15.35 is amended by revising paragraph (b) to read as follows:
Section 15.35 Measurement detector function and bandwidth.

* * * * *

(b) Unless otherwise stated, on any frequency or frequencies above 1000 MHz the radiated limits shown are based upon the use of measurement instrumentation employing an average detector function. When average radiated emission measurements are specified in the regulations, including emission measurements below 1000 MHz, there also is a limit on the radio frequency emissions, as measured using instrumentation with a peak detector function, corresponding to 20 dB above the maximum permitted average limit for the frequency being investigated unless a different peak emission limit is otherwise specified in the rules, *e.g.*, see Sections 15.255, 15.509 and 15.511. Unless otherwise specified, measurements above 1000 MHz shall be performed using a minimum resolution bandwidth of 1 MHz. Measurements of AC power line conducted emissions are performed using a CISPR quasi-peak detector, even for devices for which average radiated emission measurements are specified.

* * * * *

3. Section 15.205 is amended by adding a new subparagraph (d)(6), to read as follows:
Section 15.205 Restricted bands of operation.

* * * * *

(d)(6) Transmitters operating under the provisions of Subparts D or F of this Part.

* * * * *

4. Section 15.215 is amended by revising (c) and by removing paragraph (d), to read as follows:
Section 15.215 Additional provisions to the general radiated emission limitations.

* * * * *

(c) Intentional radiators operating under the alternative provisions to the general emission limits, as contained in Sections 15.217 *et seq.* and in Subpart E of this part, must be designed to ensure that the 20 dB bandwidth of the emission is contained within the frequency band designated in the rule section under which the equipment is operated. The requirement to contain the 20 dB bandwidth of the emission within the specified frequency band includes the effects from frequency sweeping, frequency hopping and other modulation techniques that may be employed as well as the frequency stability of the transmitter over expected variations in temperature and supply voltage. If a frequency stability is not specified in the regulations, it is recommended that the fundamental emission be kept within at least the central 80% of the permitted band in order to minimize the possibility of out-of-band operation.
5. Part 15 is amended by adding a new Subpart F, to read as follows:

Subpart F – Ultra-Wideband Operation

Section 15.501 Scope.

This subpart sets out the regulations for unlicensed ultra-wideband transmission systems.

Section 15.503 Definitions.

(a) UWB Bandwidth. For the purpose of this subpart, the UWB bandwidth is the frequency band bounded by the points that are 10 dB below the highest radiated emission, as based on the complete transmission system including the antenna. The upper boundary is designated f_H and the lower boundary is designated f_L. The frequency at which the highest radiated emission occurs is designated f_M.

(b) Center frequency. The center frequency, f_C, equals $(f_H + f_L)/2$.

(c) Fractional bandwidth. The fractional bandwidth equals $2(f_H - f_L)/(f_H + f_L)$.

(d) Ultra-wideband (UWB) transmitter. An intentional radiator that, at any point in time, has a fractional bandwidth equal to or greater than 0.20 or has a UWB bandwidth equal to or greater than 500 MHz, regardless of the fractional bandwidth.

(e) Imaging system. A general category consisting of ground penetrating radar systems, medical imaging systems, wall imaging systems through-wall imaging systems and surveillance systems. As used in this subpart, imaging systems do not include systems designed to detect the location of tags or systems used to transfer voice or data information.

(f) Ground penetrating radar (GPR) system. A field disturbance sensor that is designed to operate only when in contact with, or within one meter of, the ground for the purpose of detecting or obtaining the images of buried objects or determining the physical properties within the ground. The energy from the GPR is intentionally directed down into the ground for this purpose.

(g) Medical imaging system. A field disturbance sensor that is designed to detect the location or movement of objects within the body of a person or animal.

(h) Wall imaging system. A field disturbance sensor that is designed to detect the location of objects contained within a "wall" or to determine the physical properties within the "wall." The "wall" is a concrete structure, the side of a bridge, the wall of a mine or another physical structure that is dense enough and thick enough to absorb the majority of the signal transmitted by the imaging system. This category of equipment does not include products such as "stud locators" that are designed to locate objects behind gypsum, plaster or similar walls that are not capable of absorbing the transmitted signal.

(i) Through-wall imaging system. A field disturbance sensor that is designed to detect the location or movement of persons or objects that are located on the other side of an opaque structure such as a wall or a ceiling. This category of equipment may include products such as "stud locators" that are designed to locate objects behind gypsum, plaster or similar walls that are not thick enough or dense enough to absorb the transmitted signal

(j) Surveillance system. A field disturbance sensor used to establish a stationary RF perimeter field that is used for security purposes to detect the intrusion of persons or objects.

(k) EIRP. Equivalent isotropically radiated power, *i.e.*, the product of the power supplied to the antenna and the antenna gain in a given direction relative to an isotropic antenna. The EIRP, in terms of dBm, can be converted to a field strength, in $dB\mu V/m$ at 3 meters, by adding 95.2. As used in this subpart, EIRP refers to the highest signal strength measured in any direction and at any frequency from the UWB device, as tested in accordance with the procedures specified in Sections 15.31(a) and 15.523 of this chapter.

(l) Law enforcement, fire and emergency rescue organizations. As used in this subpart, this refers to those parties eligible to obtain a license from the FCC under the eligibility requirements specified in Section 90.20(a)(1) of this chapter.

(m) Hand held. As used in this subpart, a hand held device is a portable device, such as a lap top computer or a PDA, that is primarily hand held while being operated and that does not employ a fixed infrastructure.

Section 15.505 Cross reference.

(a) Except where specifically stated otherwise within this subpart, the provisions of Subparts A and B and of Sections 15.201 through 15.204 and Section 15.207 of Subpart C of this part apply to unlicensed UWB intentional radiators. The provisions of Sections 15.35(c) and 15.205

do not apply to devices operated under this subpart. The provisions of Footnote US 246 to the Table of Frequency Allocations contained in Section 2.106 of this chapter does not apply to devices operated under this subpart.

(b) The requirements of Subpart F apply only to the radio transmitter, *i.e.*, the intentional radiator, contained in the UWB device. Other aspects of the operation of a UWB device may be subject to requirements contained elsewhere in this chapter. In particular, a UWB device that contains digital circuitry not directly associated with the operation of the transmitter also is subject to the requirements for unintentional radiators in Subpart B of this chapter. Similarly, an associated receiver that operates (tunes) within the frequency range 30 MHz to 960 MHz is subject to the requirements in Subpart B of this chapter.

Section 15.507 Marketing of UWB equipment.
In some cases, the operation of UWB devices is limited to specific parties, *e.g.*, law enforcement, fire and rescue organizations operating under the auspices of a state or local government. The marketing of UWB devices must be directed solely to parties eligible to operate the equipment. The responsible party, as defined in Section 2.909 of this chapter, is responsible for ensuring that the equipment is marketed only to eligible parties. Marketing of the equipment in any other manner may be considered grounds for revocation of the grant of certification issued for the equipment.
Section 15.509 Technical requirements for low frequency imaging systems.

(a) The UWB bandwidth of an imaging system operating under the provisions of this Section must be below 960 MHz.

(b) Operation under the provisions of this section is limited to the following:
(1) GPRs and wall imaging systems operated by law enforcement, fire and emergency rescue organizations, by scientific research institutes, by commercial mining companies, or by construction companies.

(2) Through-wall imaging systems operated by law enforcement, fire or emergency rescue organizations.

(3) Parties operating this equipment must be eligible for licensing under the provisions of Part 90 of our rules.

(4) The operation of imaging systems under this section requires coordination, as detailed in Section 15.525 of this chapter.

(c) An imaging system shall contain a manually operated switch that causes the transmitter to cease operation within 10 seconds of being released by the operator. In addition, it is permissible to operate an imaging system by remote control provided the imaging system ceases transmission within 10 seconds of the remote switch being released by the operator.

(d) The radiated emissions at or below 960 MHz from a device operating under the provisions of this section shall not exceed the emission levels in Section 15.209 of this chapter. The radiated emissions above 960 MHz from a device operating under the provisions of this section shall not exceed the following average limits when measured using a resolution bandwidth of 1 MHz:

Frequency in MHz	EIRP in dBm
960–1,610	−65.3
1,610–1,990	−53.3
Above 1,990	−51.3

(e) In addition to the radiated emission limits specified in the above table, UWB transmitters operating under the provisions of this section shall not exceed the following average limits when measured using a resolution bandwidth of no less than 1 kHz:

Frequency in MHz	EIRP in dBm
1,164–1,240	−75.3
1,559–1,610	−75.3

(f) There is a limit on the peak level of the emissions contained within a 50 MHz bandwidth centered on the frequency at which the highest radiated emission occurs, f_M. That limit is 0 dBm EIRP. It is acceptable to employ a different resolution bandwidth, and a correspondingly different peak emission limit, following the procedures described in Section 15.521 of this chapter.

(g) Imaging systems operating under the provisions of this section shall bear the following or similar statement, as adjusted for the specific provisions in paragraph (b) of this section, in a conspicuous location on the device:

Operation of this device is restricted to law enforcement, fire and rescue officials, scientific research institutes, commercial mining companies, and construction companies. Operation by any other party is a violation of 47 U.S.C. §301 and could subject the operator to serious legal penalties.

Section 15.511 Technical requirements for mid-frequency imaging systems.

(a) The UWB bandwidth of an imaging system operating under the provisions of this section must be contained between 1990 MHz and 10,600 MHz.

(b) Operation under the provisions of this section is limited to the following:
   (1) Through-wall imaging systems operated by law enforcement, fire or emergency rescue organizations.

   (2) Fixed surveillance systems operated by law enforcement, fire or emergency rescue organizations or by manufacturers licensees, petroleum licensees or power licensees as defined in Section 90.7 of this chapter.

   (3) Parties operating under the provisions of this section must be eligible for licensing under the provisions of Part 90 of our rules.

   (4) The operation of imaging systems under this section requires coordination, as detailed in Section 15.525 of this chapter.

(c) A through-wall imaging system shall contain a manually operated switch that causes the transmitter to cease operation within 10 seconds of being released by the operator. In addition, it is permissible to operate an imaging system by remote control provided the imaging system ceases transmission within 10 seconds of the remote switch being released by the operator.

(d) The radiated emissions at or below 960 MHz from a device operating under the provisions of this section shall not exceed the emission levels in Section 15.209 of this chapter. The radiated emissions above 960 MHz from a device operating under the provisions of this section shall not exceed the following average limits when measured using a resolution bandwidth of 1 MHz:

Frequency in MHz	EIRP in dBm
960–1,610	−53.3
1,610–1,990	−51.3
1,990–10,600	−41.3
Above 10,600	−51.3

(e) In addition to the radiated emission limits specified in the above table, UWB transmitters operating under the provisions of this section shall not exceed the following average limits when measured using a resolution bandwidth of no less than 1 kHz:

Frequency in MHz	EIRP in dBm
1164–1240	−63.3
1559–1610	−63.3

(f) There is a limit on the peak level of the emissions contained within a 50 MHz bandwidth centered on the frequency at which the highest radiated emission occurs, f_M. That limit is 0 dBm EIRP. It is acceptable to employ a different resolution bandwidth, and a correspondingly different peak emission limit, following the procedures described in Section 15.521 of this chapter.

(g) Imaging systems operating under the provisions of this section shall bear the following or similar statement, as adjusted for the specific provisions in paragraph (b) of this section, in a conspicuous location on the device:

Operation of this device is restricted to law enforcement, fire and rescue officials, public utilities, and industrial entities. Operation by any other party is a violation of 47 U.S.C. §301 and could subject the operator to serious legal penalties.

Section 15.513 Technical requirements for high frequency imaging systems.

(a) The UWB bandwidth of an imaging system operating under the provisions of this section must be contained between 3100 MHz and 10,600 MHz.

(b) Operation under the provisions of this section is limited to the following:

  (1) GPRs and wall imaging systems operated by law enforcement, fire or emergency rescue organizations, by scientific research institutes, by commercial mining companies, or by construction companies.

  (2) Medical imaging systems used at the direction of, or under the supervision of, a licensed health care practitioner.

  (3) Parties operating GPRs or wall imaging systems must be eligible for licensing under the provisions of Part 90 of our rules.

  (4) The operation of imaging systems under this section requires coordination, as detailed in Section 15.525 of this chapter.

(c) An imaging system shall contain a manually operated switch that causes the transmitter to cease operation within 10 seconds of being released by the operator. In addition, it is permissible to operate an imaging system by remote control provided the imaging system ceases transmission within 10 seconds of the remote switch being released by the operator.

(d) The radiated emissions at or below 960 MHz from a device operating under the provisions of this section shall not exceed the emission levels in Section 15.209 of this chapter. The radiated emissions above 960 MHz from a device operating under the provisions of this section shall not exceed the following average limits when measured using a resolution bandwidth of 1 MHz:

Frequency in MHz	EIRP in dBm
960–1,610	−65.3
1,610–1,990	−53.3
1,990–3,100	−51.3
3,100–10,600	−41.3
Above 10,600	−51.3

(e) In addition to the radiated emission limits specified in the above table, UWB transmitters operating under the provisions of this section shall not exceed the following average limits when measured using a resolution bandwidth of no less than 1 kHz:

Frequency in MHz	EIRP in dBm
1,164–1,240	−75.3
1,559–1,610	−75.3

(f) There is a limit on the peak level of the emissions contained within a 50 MHz bandwidth centered on the frequency at which the highest radiated emission occurs, f_M. That limit is 0 dBm EIRP. It is acceptable to employ a different resolution bandwidth, and a correspondingly different peak emission limit, following the procedures described in Section 15.521 of this chapter.

(g) Imaging systems, other than medical imaging systems, operating under the provisions of this section shall bear the following or similar statement in a conspicuous location on the device:

Operation of this device is restricted to law enforcement, fire and rescue officials, scientific research institutes, commercial mining companies, and construction companies. Operation by any other party is a violation of 47 U.S.C. §301 and could subject the operator to serious legal penalties.

Section 15.515 Technical requirements for vehicular radar systems.

(a) Operation under the provisions of this section is limited to UWB field disturbance sensors mounted in terrestrial transportation vehicles. These devices shall operate only when the vehicle is operating, *e.g.*, the engine is running. Operation shall occur only upon specific activation, such as upon starting the vehicle, changing gears, or engaging a turn signal.

(b) The UWB bandwidth for a vehicular radar system operating under the provisions of this section shall be contained between 22 GHz and 29 GHz. In addition, the center frequency, f_C, and the frequency at which the highest level emission occurs, f_M, must be greater than 24.075 GHz.

(c) Following proper installation, vehicular radar systems shall attenuate any emissions within the 23.6–24.0 GHz band that appear 38 degrees or greater above the horizontal plane by 25 dB below the limit specified in paragraph (d) of this chapter. For equipment authorized, manufactured or imported on or after January 1, 2005, this level of attenuation shall be 25 dB for any emissions within the 23.6–24.0 GHz band that appear 30 degrees or greater above the horizontal plane.

For equipment authorized, manufactured or imported on or after January 1, 2010, this level of attenuation shall be 30 dB for any emissions within the 23.6–24.0 GHz band that appear 30 degrees or greater above the horizontal plane. For equipment authorized, manufactured or imported on or after January 1, 2014, this level of attenuation shall be 35 dB for any emissions within the 23.6–24.0 GHz band that appear 30 degrees or greater above the horizontal plane. This level of attenuation can be achieved through the antenna directivity, through a reduction in output power or any other means.

(d) The radiated emissions at or below 960 MHz from a device operating under the provisions of this section shall not exceed the emission levels in Section 15.209 of this chapter. The radiated emissions above 960 MHz from a device operating under the provisions of this section shall not exceed the following average limits when measured using a resolution bandwidth of 1 MHz:

Frequency in MHz	EIRP in dBm
960–1610	−75.3
1,610–22,000	−61.3
22,000–29,000	−41.3
29,000–31,000	−51.3
Above 31,000	−61.3

(e) In addition to the radiated emission limits specified in the above table, UWB transmitters operating under the provisions of this section shall not exceed the following average limits when measured using a resolution bandwidth of no less than 1 kHz:

Frequency in MHz	EIRP in dBm
1,164–1,240	−85.3
1,559–1,610	−85.3

(f) There is a limit on the peak level of the emissions contained within a 50 MHz bandwidth centered on the frequency at which the highest radiated emission occurs, f_M. That limit is 0 dBm EIRP. It is acceptable to employ a different resolution bandwidth, and a correspondingly different peak emission limit, following the procedures described in Section 15.521 of this chapter.

Section 15.517 Technical requirements for indoor UWB systems.

(a) Operation under the provisions of this section is limited to UWB transmitters employed solely for indoor operation.

  (1) Indoor UWB devices, by the nature of their design, must be capable of operation only indoors. The necessity to operate with a fixed indoor infrastructure, *e.g.*, a transmitter that must be connected to the AC power lines, may be considered sufficient to demonstrate this.

  (2) The emissions from equipment operated under this section shall not be intentionally directed outside of the building in which the equipment is located, such as through a window or a doorway, to perform an outside function, such as the detection of persons about to enter a building.

  (3) The use of outdoor mounted antennas, *e.g.*, antennas mounted on the outside of a building or on a telephone pole, or any other outdoors infrastructure is prohibited.

  (4) Field disturbance sensors installed inside of metal or underground storage tanks are considered to operate indoors provided the emissions arc directed towards the ground.

  (5) A communications system shall transmit only when the intentional radiator is sending information to an associated receiver.

(b) The UWB bandwidth of a UWB system operating under the provisions of this section must be contained between 3,100 MHz and 10,600 MHz.

(c) The radiated emissions at or below 960 MHz from a device operating under the provisions of this section shall not exceed the emission levels in Section 15.209 of this chapter. The radiated emissions above 960 MHz from a device operating under the provisions of this section shall not exceed the following average limits when measured using a resolution bandwidth of 1 MHz:

Frequency in MHz	EIRP in dBm
960–1,610	−75.3
1,610–1,990	−53.3
1,990–3,100	−51.3
3,100–10,600	−41.3
Above 10,600	−51.3

(e) In addition to the radiated emission limits specified in the above table, UWB transmitters operating under the provisions of this section shall not exceed the following average limits when measured using a resolution bandwidth of no less than 1 kHz:

Frequency in MHz	EIRP in dBm
1164–1240	−85.3
1559–1610	−85.3

(f) There is a limit on the peak level of the emissions contained within a 50 MHz bandwidth centered on the frequency at which the highest radiated emission occurs, f_M. That limit is 0 dBm EIRP. It is acceptable to employ a different resolution bandwidth, and a correspondingly different peak emission limit, following the procedures described in Section 15.521 of this chapter.

(g) UWB systems operating under the provisions of this section shall bear the following or similar statement in a conspicuous location on the device or in the instruction manual supplied with the device:

This equipment may only be operated indoors. Operation outdoors is in violation of 47 U.S.C. §301 and could subject the operator to serious legal penalties. Section 15.519 Technical requirements for hand held UWB systems.

(a) UWB devices operating under the provisions of this section must be hand held, *i.e.*, they are relatively small devices that are primarily hand held while being operated and do not employ a fixed infrastructure.

(1) A UWB device operating under the provisions of this section shall transmit only when it is sending information to an associated receiver. The UWB intentional radiator shall cease transmission within 10 seconds unless it receives an acknowledgement from

the associated receiver that its transmission is being received. An acknowledgment of reception must continued to be received by the UWB intentional radiator at least every 10 seconds or the UWB device must cease transmitting.

(2) The use of antennas mounted on outdoor structures, *e.g.*, antennas mounted on the outside of a building or on a telephone pole, or any fixed outdoors infrastructure is prohibited. Antennas may be mounted only on the hand held UWB device.

(3) UWB devices operating under the provisions of this section may operate indoors or outdoors.

(b) The UWB bandwidth of a device operating under the provisions of this Section must be contained between 3100 MHz and 10,600 MHz.

(c) The radiated emissions at or below 960 MHz from a device operating under the provisions of this section shall not exceed the emission levels in Section 15.209 of this chapter. The radiated emissions above 960 MHz from a device operating under the provisions of this section shall not exceed the following average limits when measured using a resolution bandwidth of 1 MHz:

Frequency in MHz	EIRP in dBm
960–1,610	−75.3
1,610–1,900	−63.3
1,900–3,100	−61.3
3,100–10,600	−41.3
Above 10,600	−61.3

(d) In addition to the radiated emission limits specified in the above table, UWB transmitters operating under the provisions of this section shall not exceed the following average limits when measured using a resolution bandwidth of no less than 1 kHz:

Frequency in MHz	EIRP in dBm
1,164–1,240	−85.3
1,559–1,610	−85.3

(e) There is a limit on the peak level of the emissions contained within a 50 MHz bandwidth centered on the frequency at which the highest radiated emission occurs, f_M. That limit is 0 dBm EIRP. It is acceptable to employ a different resolution bandwidth, and a correspondingly different peak emission limit, following the procedures described in Section 15.521 of this chapter.

Section 15.521 Technical requirements applicable to all UWB devices.

(a) UWB devices may not be employed for the operation of toys. Operation onboard an aircraft, a ship or a satellite is prohibited.

(b) Manufacturers and users are reminded of the provisions of Sections 15.203 and 15.204 of this chapter.

(c) As noted in Section 15.3(k) of this chapter, digital circuitry that is used only to enable the operation of a transmitter and that does not control additional functions or capabilities is not classified as a digital device. Instead, the emissions from that digital circuitry are subject to the same limits as those applicable to the transmitter. If it can be clearly demonstrated that an emission from a UWB transmitter is due solely to emissions from digital circuitry contained within the transmitter and that the emission is not intended to be radiated from the transmitter's antenna, the limits shown in Section 15.209 of this chapter shall apply to that emission rather than the limits specified in this section.

(d) Within the tables in the above rule sections, the tighter emission limit applies at the band edges. Radiated emission levels at and below 960 MHz are based on measurements employing a CISPR quasi-peak detector. Radiated emission levels above 960 MHz are based on RMS average measurements over a 1 MHz resolution bandwidth. The RMS average measurement is based on the use of a spectrum analyzer with a resolution bandwidth of 1 MHz, an RMS detector, and a 1 millisecond or less averaging time. If pulse gating is employed where the transmitter is quiescent for intervals that are long compared to the nominal pulse repetition interval, measurements shall be made with the pulse train gated on. Alternative measurement procedures may be considered by the Commission.

(e) The frequency at which the highest radiated emission occurs, f_M, must be contained within the UWB bandwidth.

(f) Imaging systems may be employed only for the type of information exchange described in their specific definitions contained in Section 15.503 of this chapter. The detection of tags or the transfer or data or voice information is not permitted under the standards for imaging systems.

(g) When a peak measurement is required, it is acceptable to use a resolution bandwidth other than the 50 MHz specified in this subpart. This resolution bandwidth shall not be lower than 1 MHz or greater than 50 MHz, and the measurement shall be centered on the frequency at which the highest radiated emission occurs, f_M. If a resolution bandwidth other than 50 MHz is employed, the peak EIRP limit shall be $20 \log(\text{RBW}/50)$ dBm where RBW is the resolution bandwidth in megahertz that is employed. This may be converted to a peak field strength level at 3 meters using $E(\text{dB}\mu\text{V/m}) = P(\text{dBmEIRP}) + 95.2$. If RBW is greater than 3 MHz, the application for certification filed with the Commission must contain a detailed description of the test procedure, calibration of the test setup, and the instrumentation employed in the testing.

(h) The highest frequency employed in Section 15.33 of this chapter to determine the frequency range over which radiated measurements are made shall be based on the center frequency, f_C, unless a higher frequency is generated within the UWB device. For measuring emission levels, the spectrum shall be investigated from the lowest frequency generated in the UWB transmitter, without going below 9 kHz, up to the frequency range shown in Section 15.33(a) of this chapter or up to $f_C + 3/(\text{pulse width in seconds})$, whichever is higher. There is no requirement to measure emissions beyond 40 GHz provided f_C is less than 10 GHz; beyond 100 GHz if f_C is at or above 10 GHz and below 30 GHz; or beyond 200 GHz if f_C is at or above 30 GHz.

(i) The prohibition in Sections 2.201(f) and 15.5(d) of this chapter against Class B (damped wave) emissions does not apply to UWB devices operating under this subpart.

(j) Responsible parties are reminded of the other standards and requirements incorporated by reference in Section 15.505 of this chapter, such as a limit on emissions conducted onto the AC power lines.

Section 15.523 Measurement procedures.

Measurements shall be made in accordance with the procedures specified by the Commission.

Section 15.525 Coordination requirements.

(a) UWB imaging systems require coordination through the FCC before the equipment may be used. The operator shall comply with any constraints on equipment usage resulting from this coordination.

(b) The users of UWB imaging devices shall supply detailed operational areas to the FCC Office of Engineering and Technology who shall coordinate this information with the Federal Government through the National Telecommunications and Information Administration. The information provided by the UWB operator shall include the name, address and other pertinent contact information of the user, the desired geographical area of operation, and the FCC ID number and other nomenclature of the UWB device. This material shall be submitted to the following address:

Frequency Coordination Branch., OET

Federal Communications Commission

445 12th Street, SW

Washington, D.C. 20554

ATTN: UWB Coordination

(c) The manufacturers, or their authorized sales agents, must inform purchasers and users of their systems of the requirement to undertake detailed coordination of operational areas with the FCC prior to the equipment being operated.

(d) Users of authorized, coordinated UWB systems may transfer them to other qualified users and to different locations upon coordination of change of ownership or location to the FCC and coordination with existing authorized operations.

(e) The NTIA/FCC coordination report shall include any needed constraints that apply to day-to-day operations. Such constraints could specify prohibited areas of operations or areas located near authorized radio stations for which additional coordination is required before operation of the UWB equipment. If additional local coordination is required, a local coordination contact will be provided.

(f) The coordination of routine UWB operations shall not take longer than 15 business days from the receipt of the coordination request by NTIA. Special temporary operations may be handled with an expedited turn-around time when circumstances warrant. The operation of UWB systems in emergency situations involving the safety of life or property may occur without coordination provided a notification procedure, similar to that contained in Section 2.405(a)–(e) of this chapter, is followed by the UWB equipment user.

Appendix B

Summary of Multipath Model for IEEE P802.15.3a

On the basis of the clustering phenomenon observed in several channel measurements, the IEEE P802.15.3a UWB channel multipath model [IEEE802 02/490] is derived from the Saleh–Valenzuela model [Saleh 1987] with slight modifications. A lognormal distribution is employed rather than a Rayleigh distribution for the multipath gain magnitude, since UWB observations show that the lognormal distribution seems to better fit the measurement data. In addition, independent fading is assumed for each cluster as well as for each ray within the cluster. Therefore, the multipath model consists of the following discrete time impulse response:

$$h_i(t) = X_i \sum_{l=0}^{L} \sum_{k=0}^{K} \alpha_{k,l}^i \delta(t - T_l^i - \tau_{k,l}^i) \tag{B.1}$$

where $\{\alpha_{k,l}^i\}$ are the multipath gain coefficients

$\{T_l^i\}$ is the delay of the lth cluster

$\{\tau_{k,l}^i\}$ is the delay of the kth multipath component relative to the lth cluster-arrival time

$\{X_i\}$ represents lognormal shadowing

i refers to the ith realization.

The model uses the definitions listed in Table B.1.

Ultra-Wideband Radio Technology   Kazimierz Siwiak and Debra McKeown
© 2004 John Wiley & Sons, Ltd   ISBN: 0-470-85931-8

Table B.1 Channel model components' definitions.

Model component	Definition
T_l	Arrival time of the first path of the lth cluster
$\tau_{k,l}$	Delay of the kth path within the lth cluster relative to the first path-arrival time T_l
Λ	Cluster-arrival rate
λ	Ray-arrival rate; the arrival rate of paths within each cluster

By definition, we have $\tau_{0,l} = 0$. The distribution of cluster-arrival time and the ray-arrival time are given by

$$p(T_l|T_{l-1}) = \Lambda \exp[-\Lambda(T_l - T_{l-1})], \quad l > 0$$

$$p(\tau_{k,l}|\tau_{(k-1),l}) = \lambda \exp[-\lambda(\tau_{k,l} - \tau_{(k-1),l})], \quad k > 0 \qquad \text{(B.2)}$$

The channel coefficients are defined as follows:

$$\alpha_{k,l} = p_{k,l}\xi_l\beta_{k,l}, \; 20 \log 10(\xi_l\beta_{k,l}) \propto \text{Normal}(\mu_{k,l}, \sigma_1^2 + \sigma_2^2),$$

$$\text{or } |\xi_l\beta_{k,l}| = 10^{(\mu_{k,l}+n_1+n_2)/20}$$

where $n_1 \propto \text{Normal}(0, \sigma_1^2)$ and $n_2 \propto \text{Normal}(0, \sigma_2^2)$ are independent and correspond to the fading on each cluster and ray respectively

$$E\left[|\xi_l\beta_{k,l}|^2\right] = \Omega_0 e^{-T_l/\Gamma}e^{-\tau_{k,l}/\gamma} \qquad \text{(B.3)}$$

where T_l is the excess delay of bin l, Ω_0 is the mean energy of the first path of the first cluster, and $p_{k,l}$ is equiprobable ± 1 to account for signal inversion due to reflections. The $\mu_{k,l}$ is given by

$$\mu_{k,l} = \frac{10\ln(\Omega_0) - 10T_l/\Gamma - 10\tau_{k,l}/\gamma}{\ln(10)} - \frac{(\sigma_1^2 + \sigma_2^2)\ln(10)}{20} \qquad \text{(B.4)}$$

In the above equations, ξ_l reflects the fading associated with the lth cluster, and $\beta_{k,l}$ corresponds to the fading associated with the kth ray of the lth cluster. Note that a complex tap model has not been adopted here. The complex base band model is a natural fit for narrowband systems to capture channel behavior independent of carrier frequency, but this motivation breaks down for UWB systems in which a real-valued simulation at RF is more natural.

Finally, since the lognormal shadowing of the total multipath energy is captured by the term X_i, the total energy contained in the terms $\{\alpha_{k,l}^i\}$ is normalized to unity for each realization. This shadowing term is characterized by the following:

$$20\log_{10}(X_i) \propto \text{Normal}(0, \sigma_x^2). \qquad \text{(B.5)}$$

Channel Characteristics Desired to Model

There are six key parameters that define the model, and these are shown in Table B.2.

These parameters are found by trying to match important characteristics of the channel. Since it is difficult to match all possible channel characteristics, the main characteristics of the channel that are used to derive the above model parameters were chosen to be the following:

- Mean excess delay

- RMS delay spread

- NP, the number of multipath components (defined as the number of multipath arrivals that are within 10 dB of the peak multipath arrival)

- Power decay profile.

Since the model parameters are difficult to match with the average power-decay profile, the main channel characteristics that are used to

Table B.2 Channel model parameters.

Model parameter	Meaning
Λ	Cluster-arrival rate
λ	Ray-arrival rate, i.e., the arrival rate of path within each cluster
Γ	Cluster-decay factor
γ	Ray-decay factor
σ_1	Standard deviation of cluster lognormal fading term (dB)
σ_2	Standard deviation of ray lognormal fading term (dB)
σ_x	Standard deviation of lognormal shadowing term for total multipath realization (dB)

Table B.3 Channel characteristics and corresponding model parameters.

Target channel characteristics [e]	CM1 [a]	CM2 [b]	CM3 [c]	CM4 [d]
Mean excess delay (ns) (τ_m)	5.05	10.38	14.18	
RMS delay (ns) (τ_{RMS})	5.28	8.03	14.28	25
$NP_{10\,dB}$			35	
NP (85%)	24	36.1	61.54	
Model Parameters				
Λ (1/ns)	0.0233	0.4	0.0667	0.0667
λ (1/ns)	2.5	0.5	2.1	2.1
Γ	7.1	5.5	14.00	24.00
γ	4.3	6.7	7.9	12
σ_1 (dB)	3.3941	3.3941	3.3941	3.3941
σ_2 (dB)	3.3941	3.3941	3.3941	3.3941
σ_x (dB)	3	3	3	3
Model Characteristics [e]				
Mean excess delay (ns) (τ_m)	4.9	9.4	13.8	26.8
RMS delay (ns) (τ_{RMS})	5	8	14	26
$NP_{10\,dB}$	13.3	18.2	25.3	41.4
NP (85%)	21.4	37.2	62.7	122.8
Channel energy mean (dB)	−0.5	0.1	0.2	0.1
Channel energy standard deviation (dB)	2.9	3.3	3.4	3.2

[a] Based on line-of-sight (LOS) (0–4 m) [IEEE802 02/240].
[b] Based on non-line-of-sight (NLOS) (0–4 m) [IEEE802 02/240].
[c] Based on of NLOS (4–10 m) [IEEE802 02/240], and NLOS in [IEEE802 02/279].
[d] Generated to fit a 25-ns RMS delay spread.
[e] Based upon a 167-ps sampling time.

determine the model parameters are the first three above. Table B.3 lists some initial model parameters for a couple of different channel characteristics that were found through measurement data.

One hundred actual realizations for each channel model are derived from the model above and are provided in the channel-model documentation [IEEE802 02/490].

Figures B.1 to B.4 show superpositions of 100 realizations of each of the channel models CM1, CM2, CM3, and CM4.

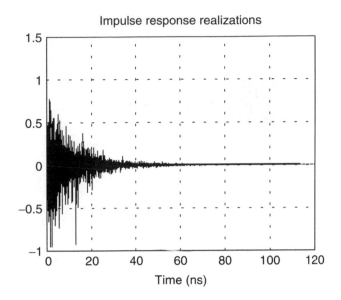

Figure B.1 Channel model CM1: 100 realizations.

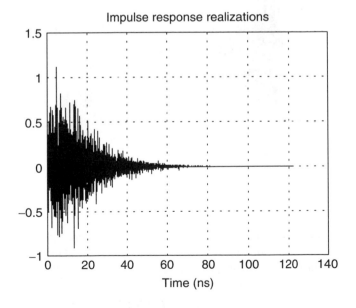

Figure B.2 Channel model CM2: 100 realizations.

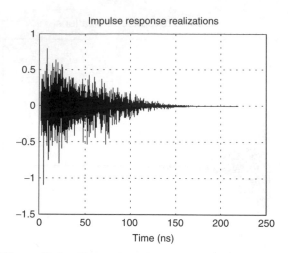

Figure B.3   Channel model CM3: 100 realizations.

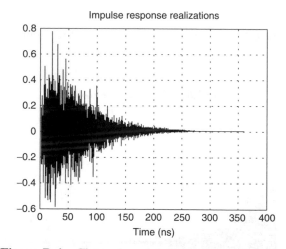

Figure B.4   Channel model CM4: 100 realizations.

References

[IEEE802 02/240] M. Pendergrass, *Empirically Based Statistical Ultra-Wideband Channel Model*, IEEE P802.15-02/240-SG3a.

[IEEE802 02/490] IEEE P802.15-02/490r0, *Channel-Modeling-Subcommittee-Report-Final-Report.doc*.

[IEEE802 02/279] J. Foerster and Q. Li, *UWB Channel Modeling Contribution from Intel*, IEEE P802.15-02/279-SG3a.

[Saleh 1987] A. Saleh and R. Valenzuela, "A statistical model for indoor multipath propagation", *IEEE JSAC*, **SAC-5**(2), 128–137, 1987.

Appendix C

Free-space Transmission of Pulses

The free-space transmission properties for impulses are derived here. In the limit of the narrowband pulse, the formulations exactly equal results for sinusoids.

Pulse Equations and Pulse Energy

We start with the current supplied by the transmitter to the antenna feed point

$$I_\text{T}(t) = (\Lambda h 2\pi a) \exp\left[\frac{-(\pi t B)^2 \log(\text{e})}{2}\right] \cos(2\pi f_\text{C}) \qquad \text{(C.1)}$$

where the 10-dB bandwidth is B. This current is supplied to an infinitesimal dipole of length Δh having radiation resistance (see [Jordan 1968, Siwiak 1998]),

$$R_\text{rad} = \frac{\eta_0}{6\pi}\left(\frac{\Delta h 2\pi f_c}{c}\right)^2 \qquad \text{(C.2)}$$

The energy W_I dissipated by the current I_T through $1\,\Omega$ is

$$W_\text{I} = \int_{-\infty}^{\infty} [I_\text{T}(t)]^2 \mathrm{d}t \qquad \text{(C.3)}$$

Ultra-Wideband Radio Technology   Kazimierz Siwiak and Debra McKeown
© 2004 John Wiley & Sons, Ltd   ISBN: 0-470-85931-8

so the transmitted pulse energy is

$$W_T = R_{\text{rad}} \int_{-\infty}^{\infty} [I_T(t)]^2 dt \tag{C.4}$$

The magnetic and electric fields due to this radiating current I_T oriented along the z-axis are

$$H_\phi(r, t) = \frac{\Delta h \sin(\theta)}{4\pi r c} \frac{\partial}{\partial t} I_z(t) \tag{C.5}$$

and

$$E_\theta = -\eta_0 H_\phi \tag{C.6}$$

The open circuit voltage at the terminals of an ideal infinitesimal receiving dipole is

$$V_R(t) = \frac{\eta_0 (\Delta h)^2}{4\pi r c} \frac{\partial}{\partial t} I_T(t) \tag{C.7}$$

The fields and received voltages require the use of a partial derivative of the current, and the energy associated with this term is

$$W_{ddt} = \int_{-\infty}^{\infty} \left(\frac{\partial}{\partial t} I_T(t) \right)^2 dt \tag{C.8}$$

For the current given by Equation (C.1),

$$W_{ddt} = \int_{-\infty}^{\infty} \left[-\exp\left[\frac{-(tB\pi)^2}{2\ln(10)} \right] \left[\frac{t \, (B\pi)^2}{\ln(10)} \cos(2\pi f_c t) \right. \right.$$
$$\left. \left. + 2\pi f_c \sin(2\pi f_c t) \right] \right]^2 dt \tag{C.9}$$

Equation (C.9) does not have a closed form solution, but can be solved numerically, or adequately approximated for bandwidths under about 150% by

$$W_{\text{app}} = (2\pi f_c)^2 W_I \tag{C.10}$$

The ratio of the exact term W_{ddt} to the approximation W_{app} is shown plotted versus the fractional bandwidth in Figure C.1.

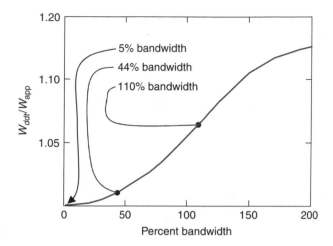

Figure C.1   Exact versus approximate time derivative of signal current pulse.

The approximation underestimates the exact value by less than 6.5% for fractional bandwidths less than 110% – the maximum possible in the 3.1- to 10.6-GHz UWB band – and by about 1% for a 44% fractional bandwidth, and essentially no error at the 5% bandwidth, so the approximation is sufficient for our purposes. As the bandwidth approaches zero, the current "pulse" becomes a cosine, and its time derivative involves the appearance of $j\omega = j2\pi f_c$, which is recognized as the derivative operator for narrowband or time-harmonic (sinusoidal) solutions to Maxwell's equations.

Energy Density and Antenna Pattern

The energy density associated with the magnetic and electric fields is found by integrating over the time duration of the pulse

$$W_{\mathrm{H}} = \eta_0 \int_{-\infty}^{\infty} [H_\phi(t)]^2 \mathrm{d}t \tag{C.11}$$

Comparing the field energy to the transmitted energy in Equation (C.4) gives us, after some algebraic manipulation,

$$\frac{W_{\mathrm{H}}}{W_{\mathrm{T}}} = 1.5[\sin(\theta)]^2 \tag{C.12}$$

which is exactly the gain pattern of an ideal infinitesimal dipole.

The Friis Transmission Formula with Constant-gain Antennas

Now we can write the expression for the energy delivered by the receiving antenna to a matched load resistance R_{rad} across the terminals of that antenna. The voltage across a matched load is half the open circuit voltage, so

$$W_R = R_{rad} \int_{-\infty}^{\infty} \left[\frac{1}{2} V_R(t) \right]^2 dt \qquad (C.13)$$

With the help of Equations (C.2), (C.4), (C.5), (C.7) and (C.10), the ratio of received to transmitted energy is found to be

$$\frac{W_R}{W_T} = \frac{c^2 G_T G_R}{(4\pi f_c r)^2} \qquad (C.14)$$

Recognizing that $G_T = G_R = 1.5$ is the gain of the ideal infinitesimal dipole at each end of the transmission link, we finally obtain a form of the Friis transmission formula (see [Friis 1946]) for energy transfer between two unity-gain or "constant-gain" antennas

$$P_L = \frac{c^2}{(4\pi f_c r)^2} \qquad (C.15)$$

When antenna gains G_T and G_R are equal to 1, the result is also recognized as the free-space propagation formula.

Constant-aperture Receive Antenna

Rather than the frequency-independent unity-gain-receiving antenna mentioned in the previous section, we can use a constant-aperture antenna. The gain G_R and antenna aperture A_e are related by the square of frequency

$$G_R = \frac{4\pi A_e f_c^2}{c^2} \qquad (C.16)$$

We can see that the Friis transmission formula of Equation (C.14) picks up another frequency dependency from (C.16). We let the transmitting antenna remain a unity-gain antenna. Propagation in free space is by itself a simple frequency-independent geometrical expansion of energy density.

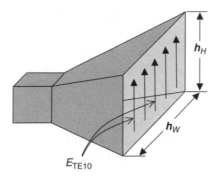

Figure C.2   The pyramidal horn, a *constant-aperture* antenna.

The *energy density* emitted by the transmitting antenna expands with distance in proportion to the area of an expanding sphere, so propagation up to the receiving antenna follows the simple geometrical expansion law $1/4\pi r^2$. The constant-gain capture area of the receiving aperture imposes the frequency dependency in Equation (C.15).

Antennas, like the pyramidal horn pictured in Figure C.2, have a *constant-aperture* behavior independent of frequency. The pyramid horn energy-capture area (operating, for example, in the TE_{10} mode) is proportional to the horn-aperture area

$$A_e = \frac{h_H h_W 8}{\pi^2} \tag{C.17}$$

and through Equation (C.16), which means that the antenna gain increases as the square of frequency:

$$G_{TE10} = \frac{32 h_H h_W f_c^2}{\pi c^2} \tag{C.18}$$

So the Friis transmission formula becomes

$$\frac{W_R}{W_{EIRP}} = \frac{A_e}{4\pi r^2} \tag{C.19}$$

or simply

$$\frac{W_R}{W_{EIRP}} = \frac{2 h_H h_W}{\pi^3 r^2} \tag{C.20}$$

There is no frequency dependency! If the transmitter pulse energy were likewise emitted by a "constant-aperture" antenna, the emitted field strength would increase in proportion to the square of the frequency, and the link between two such antennas would improve as the frequency is increased, as is evident from Equation (C.14), when both G_T and G_R are replaced by G_{TE10} for Equation (C.18). U. S. Regulatory emission limits, however, dictate the maximum effective isotropically radiated power (EIRP), so the use of gain on the transmitter side must be compensated by a corresponding reduction in the transmitter power.

References

[Friis 1946] H. T. Friis, "A note on a simple transmission formula", *Proceedings of the IRE*, **34**(5), 245–256; **41**, 1946.

[Jordan 1968] E. C. Jordan and K. G. Balmain, *Electromagnetic Waves and Radiating Systems*, Second Edition, Englewood Cliffs, NJ: Prentice-Hall, 1968.

[Siwiak 1998] K. Siwiak, *Radiowave Propagation and Antennas for Personal Communications*, Second Edition, Norwood, MA: Artech House, 1998.

Appendix D

Glossary

Definitions and Constants

Not every symbol used in the text is listed here. Some symbols are reused in the text, and these are clearly obvious in their context. Vector quantities, such as **E** and **H**, are in bold. The physical constants are from: "*2002 CODATA Recommended Values of the Fundamental Physics Constants,*" 31 December 2002, (Online): <http://physics.nist.gov/cuu/Constants/> December 2003.

Acronyms and constants:	
ADSL	Asymmetric Digital Subscriber Line
A_e	antenna aperture, m^2
ANSI	American National Standards Institute
API	application programmer's interface
AWGN	additive white Gaussian noise
B, BW	bandwidth, Hz
BER	bit error rate
bps	also b/s, bits per second
BPSK	binary phase shift keying
c	299,792,458 m/s speed, speed of light in vacuum (exact)
C	capacity, bps
CCIR	International Radio Consultative Committee
CDMA	code division multiple access

Ultra-Wideband Radio Technology   Kazimierz Siwiak and Debra McKeown
© 2004 John Wiley & Sons, Ltd   ISBN: 0-470-85931-8

Acronyms and constants:

CEPT	European Conference of Postal and Telecommunications Agencies
DAN	Device Area Network
dB	decibel
dBi	antenna gain, dB referenced to isotropic radiation
dBm	dB relative to one milliwatt
DPSK	differential phase shift keying
DSP	digital signal processor
DSSS	direct sequence spread spectrum
E_b	bit energy
e_b	communication efficiency, relative efficiency, dB
E_b/N_0	bit energy to noise-density ratio
EIRP	effective isotropically radiated power
EM	electromagnetic
ERO	European Radiocommunications Office
ETSI	European Technical Standards Institute
FDMA	frequency division multiple access
FDTD	finite difference time domain
FEC	forward error correction
FFT	Fast Fourier Transform
G	gain
h	$6.6260693 \times 10^{-34}$ Js, Planck's constant
Hz	cycles per second
IEEE	Institute of Electrical and Electronics Engineers
IFFT	Inverse Fast Fourier Transform
IP	Internet protocol
ISO	International Standards Organization
ITU	International Telecommunication Union
J	joules $=$ W s
K	kelvin, degrees above absolute zero
k_b	Boltzmann's constant, $1.3806505 \times 10^{-23}$ J/K
LAN	Local Area Network
LLC	logical link control
LOS	Line of Sight (propagation)
MAC	medium access control
MAN	Metropolitan Area Network
m_b	modulation efficiency
M-BOK	M-ary Bi-Orthogonal Keying
MBWA	mobile broadband wireless access

Acronyms and constants:	
N_0	noise density, W/Hz
NLOS	Non-line of sight (propagation)
OFDM	orthogonal frequency division multiplexing
OOK	on−off keying
PAM	pulse amplitude modulation
PAN	Personal Area Network
P_d	power density, W/m^2
PHY	physical layer
POTS	plain old telephone system
PPM	pulse position modulation
QAM	quadrature amplitude modulation
QPSK	quadrature phase shift keying
RMS	root mean square
SE	spectrum engineering
SG_{lim}	173 dB/bps, UWB system gain per bit per second; FCC, USA
SNR	also, S/N, signal-to-noise ratio
T	temperature, kelvin
TAG	Technical Advisory Group
T_B	data bit duration, s
TCP/IP	terminal control protocol/Internet protocol
TDMA	time division multiple access
TG	Technical Group
TM-UWB	time-modulated UWB
t_r	ray arrival interval
TR-UWB	transmitted reference UWB
UDP	user datagram protocol
U-NII	Unlicensed National Information Infrastructure
UWB	ultra-wideband
WAN	wide area network
WLAN	Wireless Local Area Network
λ	wavelength, m
π	3.1415926535 . . .
τ_{RMS}	delay spread, s

Index